One of the goals of this book is for as many people as possible to learn this science, so if you wish to replicate parts of the content, please be kind enough to contact me.

www.mldepot.co.uk

On the website you can find more content on machine learning. I will have two more books on this before machine learning series: Volume 2 - Calculus and Volume 3 - Probability and statistics.

If you wish to access complementary exercises as well as some coding challenges, please visit my github repository:

https://github.com/mldepot/beforeMachineLearning

Contents

Chapter 1

We Start Here

This book is a story about linear algebra, the main character of which is the star of this discipline, the vector. We will start by defining this concept, that prides itself on its simplicity. But don't mistake simplicity for lack of power; from an uncomplicated vector we will arrive at complex methods like the single value decomposition and the principal component analysis.

My journey began when I was only four years old, and my father gave me my first book on equations. Since then, I have never looked back. Mathematics flowed in my mind, and calculations came out as naturally as a delicate butterfly landing on a ravishing red petal of this miracle of nature that we so often call a flower... don't be scared already! We are just at the second paragraph, and this is not true. I am just a regular guy who was most likely kicking a ball around when he was four. But, being a typical fellow, my struggle with mathematics was real during a specific time in my life, my first couple of years at university. This was because of a combination of a bad attitude and a need for content to be structured more like a story than a manual. I was scared of equations and blamed everything I could, except myself, for my lack of success in understanding mathematics. When I look back now, I can see that it is impossible to understand anything with that attitude.

Symbols and Greek letters are the alphabets of mathematics, whereas equations are the words that represent abstract concepts. One needs to try to understand how to read this syntax, as it will bring significant benefits in the future. Unfortunately, mathematics has no sound, so I don't think you can expect good results by using a hands-on approach where you learn by ignoring the syntax, as you might do with a musical instrument. Still, as a mathematician, I can't say that this way is not possible. After all, the realm of uncertainty is where we do our best work. Once I overcame this first hurdle and I started to be able to read equations, another issue arose. I knew concepts in isolation, but relating them to one another seemed impossible. Different books have distinct structures and expose the same ideas in varying sequences, which became another obstacle for me. Now I say that I was lucky, but at the time, I considered myself the unluckiest person in the world. I could not have been more wrong.

The itinerary whereby I began putting concepts together and really understanding mathematics started on the day I missed the meeting where we, the students, were due to meet the professors who would be supervising our university theses. I can't provide a good reason for missing this meeting that won't make you think I am an idiot, but hey, sometimes things have a funny way of resolving themselves.

When I finally returned to the mathematics department, my colleagues came to me with a concerned look, enquired where I had been, and told me that I was in trouble as I had landed the worst supervisor ever. This lady was famous for being extremely demanding and challenging to get along with. On that same day, the path of my life changed completely. Indeed, she was demanding, and she presented me with a project I knew very little about, but had to master. She made me study, and did not give anything back to me unless she saw that I had made an effort. I had to go back to basics, but this time I decided to start with the most elementary concept of each subject, then I studied it in such a way that everything moving forward would have to be the result of knowledge I had previously acquired. This way, I could put everything into context.

I am still a data scientist. Well, in reality, I am a mathematician. I don't like that job title, but I also need to pay the bills. It helps me. The point is that my Master's thesis was the hardest thing I have ever done, and the conclusion is that if you make a significant effort to learn the basics, what comes afterwards will be a smoother ride. There is a lot of talk these days about the wealth gap, but I feel that another gap is emerging, one in knowledge. We like to press buttons and obsess about whatever is the next exciting thing. Modern entertainment and social media have given us all attention deficit disorder. When this is associated with a right-now mentality, it significantly contributes to this problem.

If I go back to my first experience with a mathematics book, I can understand why this might happen. With so much information out there, the minimal hurdle presented to somebody trying to learn something new is enough to make them try something else. There are a lot of us pressing buttons. Still, only a few of us are building them. If you want to succeed as a data scientist, it would be better to take a button-builder path. What this means is that you will have to learn mathematics.

I wrote this book aiming to help the reader to start and never have to look or go anywhere else for further information. There will be no need for notebooks, pens, laptops, or pencils: just the safe blueprint, a mask, and the machine gun. Oh sorry, those last items might have come from the "bank project" list... actually, you won't need much more than the Pythagorean theorem: my mistake.

Chapter 2

Why Linear Algebra?

Linear algebra is essential to forming a complete understanding of machine learning. The applications are countless, and the techniques from this discipline belong to a shared collection of algorithms widely used in artificial intelligence. Its properties and methods allow for faster computation of complex systems and the extraction of hidden relationships in sets of data.

All of this is very relevant, and it justifies the need to master this domain of mathematics. However, I think that the true power of linear algebra comes from something else. I first heard the term "beautiful" associated with mathematics while attending a lecture at university, and the word "lame" instantly sparked in my brain. I was not in love with mathematics, and I am still not.

The reality is that this science shaped my personality so much that I now understand what that professor meant when he called an equation "beautiful". That is the ability to define something complex that has consistently been proven to be true into dimensions that we can't possibly visualize, using a combination of Greek letters. Yeah, crazy.

With this comes abstraction, the most potent tool of linear algebra. It is a concept that will fuel creativity and the possibility of combining several ideas and techniques that can come from any part of knowledge acquired throughout your life, mathematical or not. Like any other ability worth learning, you have to train yourself to use it. And what better way to do so other than starting with the simplest element of linear algebra, the vector?

Chapter 3

What Is a Vector?

You can think of a vector in simple terms as a list of numbers where the position of each item in this structure matters. In machine learning, this will often be the case. For example, if you are analysing the height and weight of a class of students, in this domain, a two-dimensional vector will represent each student:

$$\vec{v} = \begin{pmatrix} v_1 \\ v_2 \end{pmatrix} = \begin{pmatrix} 1.64 \\ 64 \end{pmatrix}$$

Here v_1 represents the height of a student, and v_2 represents the weight of the same individual. Conventionally, if you were to define another vector for a different student, the position of the magnitudes should be the same. So the first element is height, followed by weight. This way of looking at vectors is often called the "computer science definition". I am not sure if that is accurate, but it is certainly a way to utilise a vector. Another way to interpret these elements that is more relevant to linear algebra is to think of a vector as an arrow with a direction dictated by its coordinates. It will have its starting point at the origin: the point $(0,0)$ in a coordinate system, such as the x, y plane. Then the numbers in the parentheses will be the vector coordinates, indicating where the arrow lands:

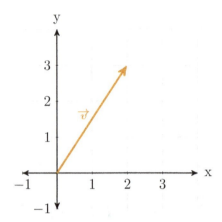

Figure 3.1: An example of a vector.

Figure 3.1 represents a visual interpretation of the vector $\vec{v}$:

$$\vec{v} = \begin{pmatrix} 2 \\ 3 \end{pmatrix}$$

As you can see, we moved from the base, point $(0,0)$, by two units to the right on the x axis and then by three units to the top on the y axis. Every pair of numbers gives you one and only one vector. We use the word pair here as we are only working with two dimensions, and the reason for using this number of coordinates is to create visual representations. However, vectors can have as many coordinates as needed. For example, if we wish to have a vector $\vec{t}$ within a three-dimensional space, we could represent it using a triplet. There is no limitation to the dimensions a vector can have.

$$\vec{t} = \begin{pmatrix} 2 \\ 3 \\ 5 \end{pmatrix}$$

In my opinion, the best way to understand linear algebra is by visualising concepts that, although simple but powerful, are often not well-understood. It is also essential to embrace abstraction, and let ourselves dive into definitions by bearing this concept in mind, as mathematics is a science of such notions.

The more you try to understand abstract concepts, the better you will get at using them. It is like everything else in life. So bear with me as I introduce the first abstract definition of the book: a vector is an object that has both a direction and a magnitude:

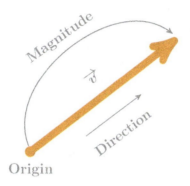

Figure 3.2: The components of a vector.

Alright, let's try and break this concept down, starting with direction. We know that a vector has a starting point, which we call the base or origin. This orientation is dependent on a set of coordinates that in itself is essential for defining this element. For example, look at the figure above. The landing point defines the direction of the vector $\vec{v}$. Magnitude is the size of the vector, and it is also a function of where it lands. Therefore, for two vectors to be the same, their directions and magnitudes must be equal. Now that we have a visual interpretation and an abstract definition, the only thing missing when it comes to fully understanding vectors is a context, a real-life situation where they are applicable. Providing context will indicate where this can be utilised and make the understanding of such notions more accessible.

For example, say that you need to describe the wind. How many information points do you need to understand the wind if you want to sail? Well, I believe that two points of information will generally be sufficient. You will certainly need the wind direction and the speed, and one vector can represent this well. If directions and magnitudes define vectors, it would be advantageous to apply mathematical operations to them; one example could be to describe paths. If we break down one course into two vectors, perhaps we could represent it as a sum of such elements.

Vectors alone are instrumental mathematical elements. However, because we define them with magnitude and directions, they can represent many things, such as gravity, velocity, acceleration, and paths. So let's hold hands and carefully make an assumption. We know nothing more about linear algebra yet, but there is a need to move on, well at least for me there is, before you call me a thief, as so far I have charged you money just for a drawing of an arrow. If we only know how to define vectors, one way to carry on moving forward will maybe be to try to combine them. We are dealing with mathematics, so let's try to add them up. But, will this make sense? Let's check.

3.1   Is That a Newborn? Vector Addition

Given two vectors $\vec{v}$ and $\vec{w}$ we can define their sum, $\vec{v} + \vec{w}$ as a translation from the edge of $\vec{v}$ with the magnitude and direction of $\vec{w}$. Did you shit yourself with all the mathematical terms? Well, there's nothing to worry about. There are two positives about that accident:

1. You can go to a party and try to impress someone with the phrase "translation from the edge of where one vector finishes of the magnitude and direction of the other one"; however, I will warn you that if somebody really is impressed by that... well take your own conclusions.

2. This is another step into the world of abstraction, and a visual explanation will follow.

Let's see what is happening here. First, let's consider two vectors: $\vec{v}$ and $\vec{w}$, which will be defined by the following pairs of coordinates:

$$\vec{v} = \begin{pmatrix} 2 \\ 3 \end{pmatrix}, \quad \vec{w} = \begin{pmatrix} 2 \\ 1 \end{pmatrix}$$

If we make use of the Cartesian plane and plot these vectors, we will have something like this:

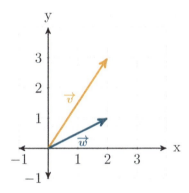

Figure 3.3: The representation of two vectors.

We want to understand geometrically what happens when we add two vectors:

$$\vec{v} + \vec{w}$$

Considering that directions and magnitudes define vectors, what would you say if you had to take a naive guess at the result of adding two vectors together? In theory, it should be another vector, but with what direction and magnitude? If I travel in the direction of $\vec{v}$ to its full extent and then turn to the direction that $\vec{w}$ points and go as far as its magnitude, I will arrive at a new location. The line from the origin to the new location where I landed, is the new vector $\vec{v} + \vec{w}$.

Let's take a pause here and analyse what just happened. Mathematics is a science developed by people, so do not become discouraged by a formula. Don't just read it, think that you can't understand it, or pretend that you did and move on. Question it!

Analytically this is how you would do this operation:

$$\vec{z} = \vec{v} + \vec{w}$$

$$= \begin{pmatrix} 2 \\ 3 \end{pmatrix} + \begin{pmatrix} 2 \\ 1 \end{pmatrix}$$

The vector $\vec{z}$ is the result of adding the elements with the same positions in vectors $\vec{v}$ and $\vec{w}$:

$$\vec{z} = \begin{pmatrix} 2+2 \\ 3+1 \end{pmatrix} = \begin{pmatrix} 4 \\ 4 \end{pmatrix}$$

We can explore a visualization to understand these so-called translations better and solidify this concept of vector addition:

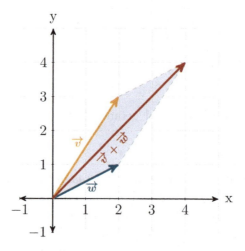

Figure 3.4: The addition of two vectors.

One can utilize vector addition in many real-life scenarios. For example, my cousin has a kid with these long arms who can throw a golf ball at 60 km/h:

Figure 3.5: My cousin's kid

One day we were driving a car north at 60 km/h. From the back seat, he threw this golf ball through the window directly to the east. If we want to comprehend the direction and velocity of the ball relative to the ground, we can use vector addition. From vector addition, we can understand that the ball will be moving north-east, and, if you wished to calculate the velocity, you could do so by using the Pythagoras theorem, $\sqrt{60^2 + 60^2}$. It is an elementary example, and the wind resistance was ignored:

Figure 3.6: My cousin's kid rock throw.

Well, if you can add vectors, it will also be possible to subtract them. It will be the same as if you were to find the sum of two vectors, but instead of going outward in the direction of the second vector, you will go the other way. There is another thing that one can do with a vector. We can change its magnitude and direction. To do so, we can multiply it by a scalar, a number that will stretch or shrink it.

3.2 Wait, Is That Vector on Steroids? Scalar Multiplication

A scalar can be any real number. If you don't recall the definition of a real number, I can refresh your memory, but what happened along the way? A real number is any number you can think of. Are you thinking about a number that seems mind blowing complicated , like π? Yeah, that is a real number. Analytically we can define a real number thus:

$$\lambda \in \mathbb{R}$$

The λ is a variable or a scalar that can accommodate any real number. The other new symbol, $\in$, means to belong to something; that something here is the set of real numbers, which we symbolise as $\mathbb{R}$. As λ is a real number, there are four different outcomes when multiplying a vector by a scalar. First, we can maintain or reverse the vector's direction depending on this scalar's sign. Another thing that will be altered is the length of the vector. If λ is between 0 and 1, the vectors will shrink, whereas if the value of λ is greater than 1 or smaller then -1, the vector will stretch:

- If $\lambda > 1$ the vector will keep the same direction but stretch.

- If $1 > \lambda > 0$ the vector will keep the same direction but shrink.

A couple more cases:

- If $\lambda < -1$ the vector will change direction and stretch.

- If $0 > \lambda > -1$ the vector will change direction and shrink.

Symbolically, multiplying a vector $\vec{v}$ by a scalar λ can be defined by:

$$\lambda \vec{v}$$

If we define a new vector $\vec{g}$ such that:

$$\vec{g} = \lambda \vec{v} \quad \text{and} \quad \lambda = -2$$

It follows:

$$\vec{g} = -2 \cdot \begin{pmatrix} 2 \\ 3 \end{pmatrix} = \begin{pmatrix} -4 \\ -6 \end{pmatrix}$$

Because we defined λ to be -2, the result of multiplying the vector $\vec{v}$ by this scalar will be a stretched version of the same $\vec{v}$ but pointing to a different direction.

In figure 3.7, we can see a visual representation of this operation.

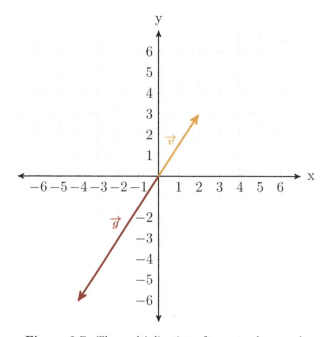

Figure 3.7: The multiplication of a vector by a scalar.

We are now at a stage where we know how to operate vectors with additions and subtractions, plus we are also capable of scaling them via multiplication with a real number. What we still haven't covered is vector multiplication.

3.3 Where Are You Looking, Vector? The Dot Product

One way that we can multiply vectors is called the dot product, which we will cover now. The other is called the cross product, which won't be covered in this book. The main difference between the dot product and the cross product is the result: the dot product result is a scalar, and what comes from the cross product is another vector. So, suppose you have two vectors of the same dimension. Then, taking the dot product between them means you will pair up the element entries from both vectors by their position, multiply them together, and add up the result of these multiplications.

So, for example, consider the vectors:

$$\vec{v} = \begin{pmatrix} 2 \\ 3 \end{pmatrix} , \vec{w} = \begin{pmatrix} 2 \\ 1 \end{pmatrix}$$

We can then calculate the dot product between $\vec{v}$ and $\vec{w}$ as such:

$$\vec{v} \cdot \vec{w} = \begin{pmatrix} 2 \\ 3 \end{pmatrix} \cdot \begin{pmatrix} 2 \\ 1 \end{pmatrix} = 2 \cdot 2 + 3 \cdot 1 = 7$$

A definition for this same concept given any two vectors, $\vec{v}$ and $\vec{w}$ where each of them have n elements, which is the same as saying that the vectors are of size n is:

$$\vec{v} \cdot \vec{w} = \sum_{i=1}^{n} v_i \cdot w_i \tag{3.1}$$

One of the goals of this series of books is for you to become comfortable with notation. I believe it will be of great value; therefore, I will describe each symbol introduced throughout the series. For example, the symbol $\sum$ is the capital Greek letter sigma; in mathematics, we use it to describe a summation, so:

$$\sum_{i=1}^{n} v_i \cdot w_i = v_1 \cdot w_1 + v_2 \cdot w_2 + ... + v_n \cdot w_n$$

In essence, we have an argument with an index, $v_i \cdot w_i$, representing what we are going to sum. Now the numbers above and below the sigma will be the controlling factors for this same index that is represented by the letter i. The number below indicates where it starts, whereas the number above is where it needs to end. In the example above, we have two components per vector and need them all for the dot product, so we have to start at one and end at two, meaning that $i = 1$ and $n = 2$.

I learned the dot product just like that. I knew I had to perform some multiplications and additions, but I couldn't understand what was happening. It was simple, and because I thought that computing it was enough, this led me to deceive myself into thinking that I fully understood it. I paid the price later, when I needed to implement these concepts to develop algorithms. The lack of context and visualisation were a killer for me.

A true understanding of linear algebra becomes more accessible with visualisations, and the dot product has a tremendous geometrical interpretation. It can be calculated by projecting the vector $\vec{w}$ into $\vec{v}$ and multiplying the magnitude of this projection with the length of $\vec{v}$, or vice versa. In other words, the dot product will represent how much of $\vec{w}$ points in the same direction as $\vec{v}$. Let's verify this; so, given the vectors $\vec{w}$ and $\vec{v}$, a projection of $\vec{w}$ into $\vec{v}$ can be represented this way:

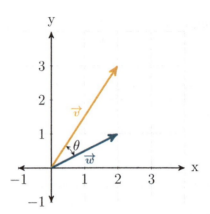

Figure 3.8: The angle between two vectors.

Projections are a fundamental concept in machine learning, particularly in understanding how data can be represented in lower-dimensional spaces. They can be intuitively understood by considering angles and movement in the context of vectors.

Imagine two vectors, say $\vec{v}$ and $\vec{w}$, originating from the same point (or having the same base). Let θ be the angle between these two vectors. This angle measures the deviation of one vector from the other in terms of direction.

However, the idea of "moving one of these vectors by θ degrees to end up on top of the other" needs clarification. In the context of vector projections, what we're actually doing is projecting one vector onto another. This projection is not about rotating the vector but rather about finding the component of one vector that lies along the direction of the other vector.

The projection of vector $\vec{w}$ onto vector $\vec{v}$ is a new vector that lies on the line of $\vec{v}$ and represents the component of $\vec{w}$ in the direction of $\vec{v}$:

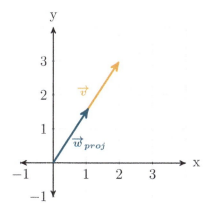

Figure 3.9: The projection of $\vec{w}$ into $\vec{v}$.

If we are after a number representing how much a vector points in the direction of another, it makes sense for us to use their lengths. Therefore, we need to calculate two measures: the length of $\vec{v}$ and the length of the projection of $\vec{w}$ into $\vec{v}$. We already know how to calculate the length of $\vec{v}$, and the norm of the projection of $\vec{w}$ into $\vec{v}$ can be derived using the Pythagorean theorem.

Given a vector's elements (the coordinates), we can quickly draw a triangle. Let's take $\vec{v}$, the vector for which we need the magnitude:

Figure 3.10: Square triangle.

So, we know that the square of the hypotenuse is equal to the sum of the squares of the cathetus (we should, right?). Generically, we can represent the vector $\vec{v}$ as $\vec{v} = (v_1, v_2)^T$, where v_1

and v_2 can take for values any real number. The vector $\vec{v}$ has been working for a long time. It is an active presence throughout math books. But, like all of us, $\vec{v}$ is ageing, and its knees are not like they used to be. So the T means that $\vec{v}$ can lie down for a moment, or in other words, it becomes transposed. Therefore T stands for a transposed version of the element in question, where columns become rows and rows become columns. We have been talking about lengths so it is about time we come up with a way of calculating such thing:

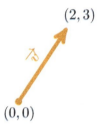

$(2, 3)$

$(0, 0)$

Figure 3.11: The length of a vector.

When we defined vectors, we established that the landing point was essential to describe the magnitude of such elements. If we have two points, the basis of the vector and the point on which the vector lands, we can use the formula to calculate the distance between two points to derive the length of this mathematical concept. It comes that we can calculate this distance with the following equation:

$$\sqrt{(x_1 - x_2)^2 + (y_1 - y_2)^2}$$

Where the x's and the y's represent the point's coordinates, if we now use the coordinates of the landing point and the origin to calculate the distance between these two points, we will have something like:

$$\sqrt{(2 - 0)^2 + (3 - 0)^2}$$

Which in turn:

$$\sqrt{2^2 + 3^2}$$

Because all vectors have the same origin, the point $(0, 0)$, the length of $\vec{v}$ is then equal to:

$$\|\vec{v}\| = \sqrt{(v_1 - 0)^2 + (v_2 - 0)^2}$$

Where v_1 and v_2 are generic coordinates for a landing point.

$$\|\vec{v}\| = \sqrt{v_1^2 + v_2^2}$$

Which gives us:

$$\|\vec{v}\| = \sqrt{2^2 + 3^2} = \sqrt{13}$$

This metric can also be called the "norm" of a vector; to be rigorous, as we should be, this norm is called the Euclidean norm. The fact that there is a name for a specific norm suggests that there are more of them in mathematics, which is exactly right. There are several definitions for norms, and each of these is calculated with a different formula. We will use the Euclidean norm for most of this book.

What if we are in a higher dimensional space, will this norm scale? The answer is yes, and we can get to the formula by using the same equation for the distance between points. So, let's try and calculate it for three dimensions, then see if we can extrapolate for n dimensions. The problem is that we are stuck with a formula that takes only two components. There are paradoxes in mathematics. Some are problems that are yet to be proven, and some are just curiosities. This one is spectacular; mathematicians do not like to do the same thing several times, so the ideal situation is to work for one case and then generalize for all. The goal is to save time so that, at the next juncture, when a similar problem arises, we have a formula that can solve it, which is an astonishing idea on paper. The funny bit is that sometimes it takes a lifetime to find the equation for the general case.

So for the next 200 pages, we will be... calm down, this one will be quite simple. So, $(v_1, v_2, v_3)^T$ can represent any vector with three coordinates. Let's use it and try to derive a generic equation for the norm formula. As it stands now, if we only consider what we have learned in this book, we can't say that we are at Einstein's level of intellect, YET! Gladly for this particular case we can get ourselves out of trouble easily:

$$\|\vec{v}\| = \sqrt{(v_1 - 0)^2 + (v_2 - 0)^2 + (v_3 - 0)^2}$$

So it comes that:

$$\|\vec{v}\| = \sqrt{v_1^2 + v_2^2 + v_3^2}$$

Generically the norm of a vector with n components can be calculated with the following equation:

$$\|\vec{v}\| = \sqrt{\sum_{i=1}^{n} v_i^2}$$

I know this is a simple concept with a rudimentary formula, but formulas can be deceiving. Sometimes, we see them, assume that they are true, move on, and think that it is understood. The problem here is that while you can just choose to trust religion, don't just trust science. Criticise it. The more you do this, the more you will learn. Blindly relying on everything that you are shown will appeal to the lazy brain as it removes the necessity to think about things. We can't forget why we did all that: the dot product! Yet another magnitude has to be calculated, but this time it is of the projection of $\vec{w}$ into $\vec{v}$. For this, we will again use trigonometry.

Yes, that word that everybody was afraid of in High School. Someone must have placed a negative connotation on that word, and I suspect it was similar to the way they handled potatoes. All of a sudden, two or three fitness blogs claim that potatoes are evil, and there you go, you can only eat sweet potatoes, or better, sweet potato fries, which are certainly much better for you than regular boiled potatoes. Figure 3.10 has a representation of a squared triangle with an angle of size θ. With the Pythagorean theorem, we can express the length of the projection of $\vec{w}$ as a function of the length $\vec{w}$ and the angle θ:

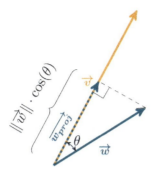

Figure 3.12: The length of the projection of $\vec{w}$ into $\vec{v}$.

Okay, so after all of this, we have a geometrical approach to the derivation of the dot product for two vectors $\vec{v}$ and $\vec{w}$ which is:

$$\vec{v} \cdot \vec{w} = \|\vec{v}\| \cdot \|\vec{w}\| \cdot \cos\theta \qquad (3.2)$$

We know from previous calculations that $\vec{v} \cdot \vec{w} = 7$. We did this when we introduced the dot product; let's verify if we get the same result with this new formula. For that, we need θ, but don't worry, I calculated it already, and it comes to around 29.4 degrees. Let's compute this bastard then:

$$\|\vec{v}\| = \sqrt{13}$$

The norm of $\vec{w}$ has the following value:

$$\|\vec{w}\| = \sqrt{5}$$

Consequently:

$$\|\vec{v}\| \cdot \|\vec{w}\| \cdot \cos\theta = \sqrt{13} \cdot \sqrt{5} \cdot \cos(29.4) = 7$$

The dot product tells you what amount of one vector goes in the direction of another (thus, it's a scalar) and hence does not have any direction.

We can have three different cases:

1. The dot product is positive, $\vec{v} \cdot \vec{w} > 0$, which means that the two vectors point in the same direction.

2. The dot product is 0, $\vec{v} \cdot \vec{w} = 0$, which means that the two vectors are perpendicular, the angle is 90 degrees.

3. The dot product is negative $\vec{v} \cdot \vec{w} < 0$, which means that the vectors point in different directions.

This may still be a bit abstract—norms, vectors, and how they align with each other's directions, so let's explore an example. Imagine we are running a streaming service where movies are represented by 2-dimensional vectors. Although this is a simplified representation, it helps us understand the applications of the dot product. In our model, each entry of our vectors represents two genres: drama and comedy. The higher the value of an entry, the more characteristics of that genre the movie has.

Our task is to recommend a movie to a user, let's call her Susan. We know that Susan has watched movie $\vec{a}$, represented by:

$$\vec{a} = \begin{pmatrix} 1 \\ 3 \end{pmatrix}$$

In our library, we have two more movies that we could recommend to Susan, movies $\vec{b}$ and $\vec{c}$:

$$\vec{b} = \begin{pmatrix} 4 \\ 2 \end{pmatrix} \quad \vec{c} = \begin{pmatrix} 2 \\ 3 \end{pmatrix}$$

Let's visualize these movie vectors :

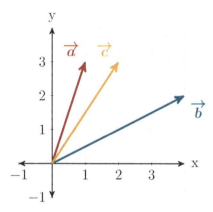

Figure 3.13: What should we recommend to Susan?

Given that we have information on Susan's movie preferences, specifically her liking for the movie represented by vector $\vec{a}$, our mission is to figure out which among the other two contenders, $\vec{b}$ and $\vec{c}$, we should recommend. It is no secret that using the dot product will make sense; this is the example for its applications, but why?

The dot product between two vectors essentially measures how much one vector aligns with the direction of the other. A higher dot product indicates a greater level of similarity. Thus, by comparing the dot products of $\vec{a}$ with both $\vec{b}$ and $\vec{c}$, we can discern which movie aligns more closely with Susan's preferences. The movie corresponding to the vector with the highest dot product emerges as the most similar to $\vec{a}$, the most likely candidate for recommendation.

However, we've already established how to calculate the dot product, so why don't we take the opportunity to explore another widely used method in machine learning? One that builds on this same concept is the cosine similarity.

Recall that the dot product is expressed by 3.2, where θ is the angle between vectors $\vec{v}$ and $\vec{w}$. Cosine similarity focuses on this $\cos(\theta)$ term, revealing how vectors are oriented concerning each other.

To compute cosine similarity, we rearrange the dot product formula to isolate $\cos(\theta)$:

$$\cos\theta = \frac{\vec{v} \cdot \vec{w}}{\|\vec{v}\| \cdot \|\vec{w}\|} \tag{3.3}$$

We have a phew things to compute. Let's start with the norms:

$$\|\vec{a}\| = \sqrt{1^2 + 3^2} = \sqrt{1 + 9} = \sqrt{10}$$

$$\|\vec{b}\| = \sqrt{4^2 + 2^2} = \sqrt{16 + 4} = \sqrt{20}$$

$$\|\vec{c}\| = \sqrt{2^2 + 3^2} = \sqrt{4 + 9} = \sqrt{13}$$

There we go! We are just missing two dot products:

$$a \cdot b = 1 \cdot 4 + 3 \cdot 2 = 4 + 6 = 10$$

$$a \cdot c = 1 \cdot 2 + 3 \cdot 3 = 2 + 9 = 11$$

Now for the cosines:

$$\cos(\theta)_{a,b} = \frac{\vec{a} \cdot \vec{b}}{\|\vec{a}\| \cdot \|\vec{b}\|} = \frac{10}{\sqrt{10} \cdot \sqrt{20}} = \frac{10}{\sqrt{200}} \approx 0.707$$

$$\cos(\theta)_{a,c} = \frac{\vec{a} \cdot \vec{c}}{\|\vec{a}\| \cdot \|\vec{c}\|} = \frac{11}{\sqrt{10} \cdot \sqrt{13}} = \frac{11}{\sqrt{130}} \approx 0.964$$

After calculating the cosines, we observe that its value between $\vec{a}$ and $\vec{b}$ is approximately 0.707, while the other cosine between $\vec{a}$ and $\vec{c}$ is approximately 0.964. It's essential to understand the behavior of the cosine in this context: this function ranges from -1 to 1, where 1 indicates that the vectors are perfectly aligned (pointing in the same direction), 0 shows orthogonality (no similarity), and -1 means that the vectors are opposed.

In our case, a higher cosine value signifies greater similarity. Therefore, the movie represented by vector $\vec{c}$, with a cosine of

0.964 concerning Susan's watched movie $\vec{a}$, shows a higher degree of similarity compared to the movie represented by vector $\vec{b}$, which has a cosine of 0.707. This implies that $\vec{c}$ shares more characteristics with Susan's preferred genres, so please enjoy your evening with movie $\vec{c}$.

Visually, we could observe this result as this is a straightforward example. If both methods were applicable here, meaning we could have used the dot product or the cosine similarity, which one should we pick?

The choice between using the dot product and cosine similarity depends on the specific goals of your analysis. Use the dot product when the magnitude of vectors is crucial, as it provides a measure that combines both direction and magnitude. In contrast, opt for cosine similarity when you're interested in comparing the direction or orientation of vectors regardless of their size. This makes cosine similarity ideal for applications like text analysis or similarity measurements in machine learning, where the relative orientation of vectors is more important than their length.

So, we now have a few ways to manipulate vectors. With this comes power and consequently, responsibility, because we need to keep these bad boys within certain limits.

3.4 A Simpler Version of the Universe - The Vector Space

In linear algebra, a vector space is a fundamental concept that provides a formal framework for the study and manipulation of vectors. Vectors are entities that can be added together and multiplied by scalars (numbers from a given field such as $\mathbb{R}$), and vector spaces are the settings where these operations take place in a structured and consistent manner.

A vector space can be thought of as a structured set, where

this set is not just a random collection of objects but a carefully defined group of entities, known as vectors, which adhere to specific rules. In mathematical terms, a vector space is a set equipped with two operations, forming a triple $(O, +, *)$.

If the word set is new to you, you can think of it as a collection of unique items, much like a pallet of colors. Each color in the pallet is different, and you can have as many colors as you like. In the world of math, a set is similar: it's a group of distinct elements, like numbers or objects, where each element is unique and there's no particular order to them. Just like how our color pallet might have a variety of colors, a set can contain a variety of elements.

In linear algebra, when we talk about a vector space, we're not just referring to a simple set like our pallet. Instead, we describe it using a special trio: $(O, +, *)$. Think of this trio as a more advanced version of our pallet, where we not only have a collection of colors (or elements) but also ways to mix and blend them to create new tones. Meaning if we blend two colors (addition) we will still have a color. Or if we add more of the same color, we still have a color (multiplication by a scalar). So if we define O such that elements are vectors:

- O is a non-empty set whose elements are called vectors.

- $+$ is a binary operation (vector addition) that takes two vectors from O and produces another vector in O.

- $*$ is an operation (scalar multiplication) that takes a scalar (from a field, typically $\mathbb{R}$ or $\mathbb{C}$) and a vector from O and produces another vector in O.

We mentioned rules but what are those ?

- if $\vec{v}$ and $\vec{w}$ are two vectors $\in O$ then $\vec{v} + \vec{w}$ must be in O.

- $\lambda \in \mathbb{R}$ and $\vec{v} \in O$ then $\lambda \vec{v}$ also needs to be in O.

Let's not refer to what follows as rules; let's call them axioms, and they must verify certain conditions to uphold the structure of a vector space. Axioms are fundamental principles that don't require proof within their system but are essential for defining and maintaining the integrity of that system. In the case of vector spaces, these axioms provide the necessary framework to ensure that vectors interact in consistent and predictable ways. Each axiom must hold true for a set of vectors to be considered a vector space. This means that for any collection of vectors to qualify as a vector space, they must adhere to these axioms, ensuring operations like vector addition and scalar multiplication behave as expected.

- Commutative property of addition:

$$\vec{v} + \vec{w} = \vec{w} + \vec{v}$$

Don't worry there are a few more:

- Associative property of addition:

$$\vec{v} + (\vec{w} + \vec{t}) = (\vec{v} + \vec{w}) + \vec{t}$$

- A zero vector exists:

$$\vec{v} + 0 = \vec{v}$$

- An inverse element exists:

$$\vec{v} + (\vec{-v}) = \vec{0}$$

Take two more to add to the collection, and these ones are

for free!

- Scalars can be distributed across the members of an addition:
$$c(\vec{v} + \vec{w}) = c\vec{v} + c\vec{w}$$

- Just as an element can be distributed to an addition of two scalars:

$$(c + d)\vec{v} = c\vec{v} + d\vec{v}$$

The last two!

- The product of two scalars and an element is equivalent to one of the scalars being multiplied by the product of the other scalar and the element:

$$(cd)\vec{v} = c(d\vec{v})$$

- Multiplying an element by 1 just returns the same element:
$$1 \cdot \vec{v} = \vec{v}$$

Do you need to know these axioms to apply machine learning? Well, not really. We all take them for granted. The reason that I have included them is to try and spark some curiosity in your mind. We all take things for granted and fail to appreciate and understand the things that surround us. Have you ever stopped to think about what happens when you press a switch to turn a light bulb on? Years of development and studying had to take place for that object to be transformed into a color that illuminates your surroundings.

I feel that this is the right moment for an example, so let's check out one of a vector space and another of a non-vector space. For a vector space, let's consider $\mathbb{R}^2$. This is the space formed by all of the vectors with two dimensions, whose elements are real numbers. Firstly, we need to verify whether we will still end up with two-dimensional vectors with real entries after adding any two vectors. Then, we want to check that we obtain a new stretched or shrunk version of a vector that is still is in $\mathbb{R}^2$ after we multiply the vector by a scalar. As there is a need to generalize, let's define two vectors $\overrightarrow{v}$ and $\overrightarrow{w}$ as being in $\mathbb{R}^2$. Let's also define a scalar $\lambda \in \mathbb{R}$. If we multiply $\overrightarrow{v}$ by λ we have:

$$\lambda \cdot \overrightarrow{v} = \begin{pmatrix} \lambda \cdot v_1 \\ \lambda \cdot v_2 \end{pmatrix}$$

So, $\lambda \cdot \overrightarrow{v}$ has size two and both $\lambda \cdot v_1$, $\lambda \cdot v_2$ are real numbers because multiplying a real number by a real number results in a new real number. The only bit we are missing is the verification of what happens when we add $\overrightarrow{v}$ with $\overrightarrow{w}$:

$$\overrightarrow{v} + \overrightarrow{w} = \begin{pmatrix} v_1 + w_1 \\ v_2 + w_2 \end{pmatrix}$$

We have a vector of size two, and its elements belong to $\mathbb{R}$, as the addition of real numbers produces a real number, therefore $\mathbb{R}^2$ is a valid vector space (all the axioms also verify, the proofs are similar so will be omitted). For the example of a non-valid vector space, consider the following, $\mathbb{R}^2_{\neq(0,0)}$. This is the set of vectors with two dimensions excluding $(0,0)^T$. This time we have:

$$\overrightarrow{v} = \begin{pmatrix} 1 \\ 1 \end{pmatrix} \quad \text{and} \quad \overrightarrow{w} = \begin{pmatrix} -1 \\ -1 \end{pmatrix}$$

Adding these vectors results in $(0,0)^T$, which is not part of $\mathbb{R}^2_{\neq(0,0)}$.

Since we have spoken so much about $\mathbb{R}^2$, what would be really helpful is if there was a way to represent all of the vectors belonging to such a space. And, as luck would have it, there is such a concept in linear algebra called linear combination; this mechanism allows the derivation of new vectors via a combination of others. The term linear suggests that there are no curves, just lines, planes, or hyper-planes, depending on the dimensions that we are working in. Here we are working with two dimensions, and a linear combination can be, for example:

$$\alpha \cdot \begin{pmatrix} 1 \\ 0 \end{pmatrix} + \beta \cdot \begin{pmatrix} 0 \\ 1 \end{pmatrix} \quad \text{where} \quad \alpha, \beta \in \mathbb{R}$$

Say that we name $(1,0)^T$ as $\vec{i}$ and the vector with coordinates $(0,1)^T$ as $\vec{j}$. Let's plot these vectors alongside the vector, $\vec{v}$:

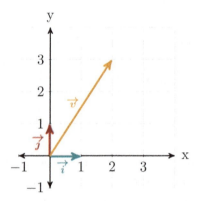

Figure 3.14: A vector as a linear combination.

The vector $\vec{v}$ has for coordinates $(2,3)^T$, which means that if we stretch $\vec{i}$ by two units and then sum it to a three units stretched version of $\vec{j}$, the result will be equal to $\vec{v}$. Let's call i^* to the stretched version of $\vec{i}$ and j^* to the stretched version of $\vec{j}$.

Algebraically we can represent $\vec{v}$ as:

$$\vec{v} = 2 \cdot \begin{pmatrix} 1 \\ 0 \end{pmatrix} + 3 \cdot \begin{pmatrix} 0 \\ 1 \end{pmatrix}$$

Which is the same as:

$$\vec{v} = 2 \cdot \vec{i} + 3 \cdot \vec{j}$$

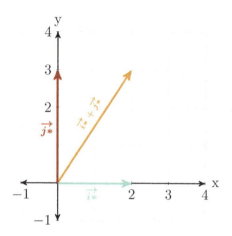

Figure 3.15: A new representation of $\vec{v}$ as a linear combination.

If we now replace the scalars two and three with two variables, which we'll call α and β, where both of them are in $\mathbb{R}$, we get:

$$\alpha \cdot \vec{i} + \beta \cdot \vec{j} \qquad (3.4)$$

We can display all the vectors of the vector space $\mathbb{R}^2$ using equation 3.4. Let's think about this affirmation for a second and see if it makes sense. If I have the entire set of real numbers assigned to the scalars α and β, it means that if I add up the scaled version of $\vec{i}$ and $\vec{j}$, I can get any vector within $\mathbb{R}^2$.

The vectors $\vec{i}$ and $\vec{j}$ have a particular property that is important to mention; they are linear independent. What this means is that you can't get to $\vec{j}$ via $\vec{i}$ and vice versa. Mathematically this is defined by the following equation:

$$c_1 \cdot \vec{v_1} + c_2 \cdot \vec{v_3} + ... + c_n \cdot \vec{v_n} = 0 \qquad (3.5)$$

In the equation 3.5, the factors $c_1, c_2, ..., c_n$ are scalars or real numbers. The $v's$ are a set of vectors that belong to the space and are linearly independent if, and only if, the values for the $c's$ that satisfy that equality are 0. Let's verify if 3.4 satisfies this property:

$$\alpha \cdot \begin{pmatrix} 1 \\ 0 \end{pmatrix} + \beta \cdot \begin{pmatrix} 0 \\ 1 \end{pmatrix} = 0$$

The only way for the equality to be true is if both α and β are equal to zero. Therefore, $\vec{i}$ and $\vec{j}$ are linearly independent.

So, if the condition for linear independence is that the scalar values represented by the $c's$ have to be zero, then the opposite, meaning that at least one of them is not equal to zero, must mean that the vectors are linearly dependent, for example:

$$c_1 \cdot \vec{v_1} = -c_2 \cdot \vec{v_2}$$

Now, say that instead of $(1,0)^T$ and $(0,1)^T$ we have $\vec{w} = (1,1)^T$ and $\vec{z} = (2,2)^T$:

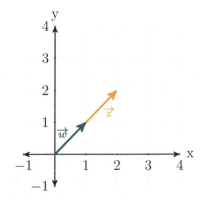

Figure 3.16: The showcase of linearly dependent vectors.

If we define a linear combination of these two vectors, this is what it will look like:

$$\alpha \cdot \begin{pmatrix} 1 \\ 1 \end{pmatrix} + \beta \cdot \begin{pmatrix} 2 \\ 2 \end{pmatrix} \tag{3.6}$$

For any values of α and β, all of the resultant vectors from this linear combination 3.6 will land on the same line:

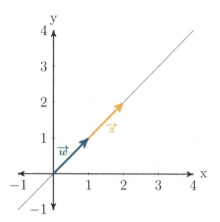

Figure 3.17: A line formed by the linearly dependent vectors.

This happens because $\vec{w}$ and $\vec{z}$ are not linearly independent. So, we can get to $\vec{w}$ by $\vec{z}$ and vice versa:

$$\vec{z} = 2 \cdot \vec{w} \tag{3.7}$$

What we can observe from equation 3.6 is that we are not able to represent .all of the vectors in the space using the two vectors, $\vec{z}$ and $\vec{w}$. Another thing we can learn from this experiment is the concept of span; the continuing black line is an example of a span. All of the vectors that result from a linear combination define the span. For instance, in the case of $\vec{i}$ and $\vec{j}$, the span is the entire vector space because we can get all the vectors within the vector space with a linear combination of these vectors, whereas with $\vec{z}$ and $\vec{w}$, the span is a line. With this, we have arrived at the definition of a basis; for a set of vectors to be considered a basis of a vector space, these vectors need to be linearly independent, and their span has to be equal to the entire vector space, therefore $\vec{i}$ and $\vec{j}$ form a basis of $\mathbb{R}^2$. A vector space can have more than one basis, and $\mathbb{R}^2$ is one example of a space that contains many basis.

When I first read this in a book, I was more confused than a horse riding a human. Why is there a need for more than one basis? I always felt tall. When frequenting bars, my shoulders were usually above the same pieces of anatomy of most people in my surroundings. One day, I went to the Netherlands, and boy oh boy, I felt short. I could see shoulders everywhere! My height had not changed, but my perspective had. You can think of a basis in the same way, perspectives from which we observe the same vector in different ways. Let's define three vectors, two to form a basis, which will be $\vec{i} = (1,0)^T$ and $\vec{j} = (0,1)^T$, the standard basis. And let's consider another vector so we can understand what happens to it when the basis is changed, so let's put the vector $\vec{v} = (2,3)^T$ back to work:

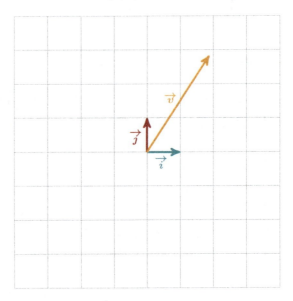

Figure 3.18: The vector $\vec{v}$ from the perspective of the standard basis.

The grids represent our perspective or the basis, which is the way we observe $\vec{v}$ on the basis formed by $\vec{i}$ and $\vec{j}$. If we wish to write $\vec{v}$ via a linear combination, we can stretch the coordinate x by two units and the second coordinate, y, will be scaled by three units:

$$\vec{v} = 2 \cdot \vec{i} + 3 \cdot \vec{j}$$

Cool, more of the same stuff, a vector on an x,y axes where x is perpendicular to y. Say that we derive a new basis, a set of two linearly independent vectors whose span is the vector space, for example, $\vec{w} = (1,0)^T$ and $\vec{z} = (1,1)^T$. These vectors are no longer perpendicular. The grid in the previous plot will change, and, consequently, so will the perspective from which we observe $\vec{v}$:

Figure 3.19: The vector $\vec{v}$ from a perspective of a different basis.

If we wish to calculate the new coordinates of $\vec{v}$ using this new basis, we can once again make use of a linear combination:

$$\vec{v} = v_1^* \cdot \vec{w} + v_2^* \cdot \vec{z}$$

Where, v_1^* and v_2^* are the coordinates of $\vec{v}$ in the new basis. That expression becomes:

$$\begin{pmatrix} 2 \\ 3 \end{pmatrix} = v_1^* \cdot \begin{pmatrix} 1 \\ 0 \end{pmatrix} + v_2^* \cdot \begin{pmatrix} 1 \\ 1 \end{pmatrix}$$

This will result in two equations:

$$2 = v_1^* + v_2^*$$

$$v_2^* = 3$$

By replacing the value of v_2^* in the first equation we get that the coordinates of $\vec{v}$ in the new basis formed by $\vec{w}$ and $\vec{z}$ are $(-1,3)^T$.

Understanding both the concept of a basis and the result of changing it is fundamental. As we advance, these manipulations will be a recurrent theme, and not only will we change the basis inside the same space, but we will also change spaces. Techniques like this will be useful because you can find properties of vectors or data using a different basis or spaces, which allows for faster computation or even better results when dealing with machine learning models.

For example, say that we wish to predict the price of a house, and our dataset consists of two measurements: the number of bedrooms and the number of bathrooms. In this context, the vector $\vec{i} = (1,0)^T$ will point in the directions where the number of bedrooms increases, whereas the number of bathrooms follows the orientation of the vector $\vec{j} = (0,1)^T$. Suppose now that, after looking into the data, we noticed that houses with more rooms in total tend to have higher prices. Another trend we might have captured was that when the number of bathrooms is the same or close to the number of bedrooms, the higher price phenomenon is also present.

Given this scenario, we can use a new basis to get a different perspective on the data, one that will allow us to understand those trends better. I have an idea, what if we defined $\vec{i^*} = (1,1)^T$ and $\vec{j^*} = (1,-1)^T$ as the new basis? Let's think about this one for a second. The vector $\vec{i^*}$ will represent the total number of rooms. On the other hand, $\vec{j^*}$ displays the difference in number between bedrooms and bathrooms.

So, in this new basis, the house's features are expressed not in terms of the absolute numbers of bedrooms and bathrooms but in terms of its total number of rooms (the first basis vector) and how balanced the number of bedrooms and bathrooms are (the second basis vector).

Should we change something? What about a basis? Sure thing! Say that we have the vector $\vec{w}$ defined as such:

$$\vec{w} = \begin{pmatrix} 3 \\ 2 \end{pmatrix}$$

This vector represents a house that is somewhat balanced in terms of rooms. It has three bedrooms and two bathrooms. Let's check it from a new perspective, the one described by the new basis, $\overrightarrow{i^*}$ and $\overrightarrow{j^*}$:

$$\begin{pmatrix} 3 \\ 2 \end{pmatrix} = w_1^* \cdot \begin{pmatrix} 1 \\ 1 \end{pmatrix} + w_2^* \cdot \begin{pmatrix} 1 \\ -1 \end{pmatrix}$$

OK, so we know that $3 = w_1^* + w_2^*$ and $2 = w_1^* - w_2^*$ meaning that:

$$w_2^* = 3 - w_1^*$$

Therefore:

$$2 = w_1^* - (3 - w_1^*) \Rightarrow w_1^* = \frac{5}{2}$$

So:

$$w_2^* = \frac{1}{2}$$

The vector $\overrightarrow{w^*} = (\frac{5}{2}, \frac{1}{2})^T$ represents the same house but in different terms. The first component is associated with the vector $\overrightarrow{i^*} = (1, 1)^T$, representing a direction in which the number of bedrooms and bathrooms increases equally. So, a value of $\frac{5}{2}$ in this direction suggests that the house has a somewhat balanced distribution of bedrooms and bathrooms. The second component is associated with the other vector that constitutes the new basis, $\overrightarrow{j^*} = (1, -1)^T$. Here, $\frac{1}{2}$ suggests that the house has more bedrooms than bathrooms by half a room's worth.

This change in basis transformed the house's representation that focused on the raw number of bedrooms and bathrooms, to the new basis, where attention defers to the balance and imbalance of the number of rooms. As a result, it's a valuable technique for revealing different aspects of the data or preparing it for further analysis or manipulation. For example, it might be easier to analyse or predict house prices regarding the balance and imbalance of the number of rooms rather than the raw numbers of bedrooms and bathrooms. A great example of an algorithm that uses such a technique is the principal component analysis, which we will cover in the last chapter.

Needless to say, there is another mathematical way to perform these transformations. For this purpose, we will be making use of something that is probably familiar to you, matrices.

Chapter 4

But What About a Matrix?

A matrix is a rectangular arrangement of numbers, symbols, or functions organized in rows and columns. In various contexts, a matrix may denote a mathematical object or an attribute of such an object. A key feature of these elements is their order or size, defined by the number of rows and columns. Matrices are commonly denoted by capital letters, and can be represented using the following notation:

$$A_{m \times n}$$

In this representation, m is the number of rows and n is the number of columns. An example of a 3×3 matrix is:

$$A_{3 \times 3} = \begin{pmatrix} 1 & 4 & 12 \\ 9 & 5 & 1 \\ 8 & 2 & 7 \end{pmatrix}$$

If we wish to generalise a representation for all matrices, we can rewrite the one above in a form that will accommodate any case:

$$A_{m \times n} = \begin{pmatrix} a_{11} & a_{12} & \cdots & a_{1n} \\ a_{21} & a_{22} & \cdots & a_{2n} \\ \vdots & \vdots & \ddots & \vdots \\ a_{m1} & a_{m2} & \cdots & a_{mn} \end{pmatrix}$$

The $a's$ are referred to as an entry, with each of these being defined by their index. For instance, a_{11} is the element in the first row and the first column of the matrix. In machine learning, the use of matrices mainly accommodates three different cases. Let me explain.

4.1 When a Table Isn't Furniture

There is one element without which we would be unable to do any work with machine learning algorithms; features. Features are what characterise your data. They are a set of metrics per element of granularity, from which your algorithm will try to learn signals. This same set can be represented as a vector, and, if you combine all the vectors into a single structure, you end up with a matrix. For example, say that we wish to predict whether a student will make it onto the basketball team based on their weight, height, and an average of their grades (assuming that their grades are a real number). Here the granularity, a word that reflects the level of detail present in the data, is the students, and the metrics are height, weight, and the average grade. Students can be represented by vectors:

$$
\begin{aligned}
\text{Student}_1 &= (1.65, 64, 5) \\
\text{Student}_2 &= (1.54, 67, 4) \\
&\vdots \\
\text{Student}_n &= (1.61, 55, 2)
\end{aligned}
$$

Combining these vectors will produce a feature space matrix that can be placed alongside a target variable. An example of such a variable is a Boolean indicator; that's just a fancy name for a variable that can only take two values, for example, yes or no. This variable will then exhibit whether or not a student made it onto the basketball team. This enriched data set can then be used to perform predictions:

$$\begin{pmatrix} 1.65 & 64 & 5 & yes \\ 1.54 & 67 & 4 & yes \\ \vdots & \vdots & \vdots & \vdots \\ 1.61 & 55 & 2 & no \end{pmatrix}$$

4.2 Yoga for Vectors - Linear Transformations

The second representation of a matrix is a linear transformation. Bear with me; this is a concept that has the potential to change your learning experience of linear algebra. As this notion is related to movement in space, we can showcase it using visualisations. This will help to solidify a deep understanding of an idea that, although simple, is essential when it comes to understanding everything else in this book. It is an excellent tool to have in your arsenal, and it includes most of the techniques already studied.

Breaking down the term will give a pretty good indication of what we are dealing with. In fact, this is good practice in mathematics. After all, the names have to make some kind of sense. Actually, for a long time these names did not make sense to me because I always associated them with a formula, and never stopped to think about what the names truly meant. This cost me time, but for some reason, one random day, it clicked. I remember with which one it was, the standard deviation. I was walking along, and all of a sudden, almost like an epiphany, everything was clear in my mind. Standard deviation is how much data deviates from the standard. But what is the standard? Probably the mean. Following the happiness of this moment, one of the best sequences of letters (that works like a charm in the right situation) came to me out loud; Motherfucker!

Now, we have covered the meaning of the word linear: no curves please, so all lines will remain lines. Transformation is another word for mapping or function. Also, is it important to mention that, despite the tasks in hand, the origin of the vectors must remain intact in transformations. So we will be transforming vectors into vectors or scalars via matrices. Analytically, a linear transformation L between two spaces Z and R can be represented as:

$$Z \overset{L}{\to} R$$

The vector is going from Z to R via L. More rigorously, a linear transformation L is defined as:

$$L : Z \to R$$

Since we are not gonna be considering any complex numbers in this book, we can even be a bit more particular and define the linear transformations as:

$$L : \mathbb{R}^n \to \mathbb{R}^m$$

This means we will move a vector from a space with n dimensions into a space with m dimensions. For the purpose of demonstration, let's consider $n = m = 2$. I can give you an ego booster here; when $n = m$, the linear transformation is called an endomorphism. Don't say I am not your friend, dropping all these fancy names.

We are defining a way to move a vector in space via patterns that allow it to keep its origin, but also to rotate and/or stretch. The transformation defined by L will map every vector in the same way. In the following example, L is a 180 degree rotation of any vector from the vector space $\mathbb{R}^2$:

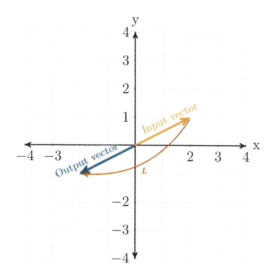

Figure 4.1: An example of a linear transformation.

The plot bellow, is to showcase the fact that a linear transformation will do the same thing to every vector that we decide to map:

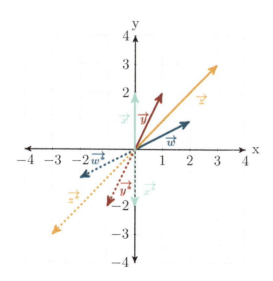

Figure 4.2: Some mappings via the same linear transformation.

But What About a Matrix?

This transformation is defined by:

$$L : \mathbb{R}^2 \to \mathbb{R}^2$$

$$(l_1, l_2) \to (-l_1, -l_2)$$

The letters, l_1, l_2 represent generic coordinates of a given vector, for example, given any vector $\vec{v} \in \mathbb{R}^2$, with coordinates (v_1, v_2) the linear transformation L will transform the coordinates of this vector into their negations, $(-v_1, -v_2)$. Negation is not a depreciative term. It means the negative version of a positive number, or vice versa, and numbers do not have feelings, well, so far. Analytically, we have defined linear transformations, but we have yet to understand why we need matrices for these functions. To do so, let's start by considering one basis of $\mathbb{R}^2$, the one defined by $\vec{i}$ and $\vec{j}$ where:

$$\vec{i} = \begin{pmatrix} 1 \\ 0 \end{pmatrix} \quad \text{and} \quad \vec{j} = \begin{pmatrix} 0 \\ 1 \end{pmatrix}$$

So any vector that belongs to $\mathbb{R}^2$ can be represented as:

$$\alpha \cdot \vec{i} + \beta \cdot \vec{j}$$

Which is the same as having:

$$\alpha \cdot \begin{pmatrix} 1 \\ 0 \end{pmatrix} + \beta \cdot \begin{pmatrix} 0 \\ 1 \end{pmatrix} \tag{4.1}$$

Where $\alpha, \beta \in \mathbb{R}$. If you don't believe me, I will dare you to find a vector that belongs to $\mathbb{R}^2$ that can't be obtained by equation 4.1. Let's make this interesting, I will buy a car for the person that finds out about such a vector. Subscribe to the mailing list to participate in this context. (Note: *This is a marketing trick, don't forget to erase this note before printing the book*). Now, let's grab a vector $\vec{v} \in \mathbb{R}^2$ and write it down as a linear combination of $\vec{i}$ and $\vec{j}$. If we select the values $(2, 3)^T$ as the coordinates of $\vec{v}$, it follows that:

$$\vec{v} = 2 \cdot \vec{i} + 3 \cdot \vec{j} \tag{4.2}$$

We defined L on the fact it will rotate any vector in $\mathbb{R}^2$ by 180 degrees. If we apply this transformation to the basis $\vec{i}$, $\vec{j}$ and call the resultant vectors as $\vec{i^*}$, $\vec{j}*$ we have:

$$\vec{i^*} = \begin{pmatrix} -1 \\ 0 \end{pmatrix} \quad \text{and} \quad \vec{j^*} = \begin{pmatrix} 0 \\ -1 \end{pmatrix}$$

Alright, let's grab equation 4.2 and replace $\vec{i}$ and $\vec{j}$ by $\vec{i^*}$ and $\vec{j^*}$ respectively:

$$\vec{v^*} = 2 \cdot \vec{i^*} + 3 \cdot \vec{j^*}$$

Here, $\vec{v^*}$ represents a new version of $\vec{v}$ which is defined as a linear combination of the transformed base. It follows that:

$$\vec{v^*} = 2 \cdot \begin{pmatrix} -1 \\ 0 \end{pmatrix} + 3 \cdot \begin{pmatrix} 0 \\ -1 \end{pmatrix} = \begin{pmatrix} -2 \\ -3 \end{pmatrix}$$

Well, $\vec{v^*}$ is the same as $-\vec{v}$, meaning that $\vec{v^*}$ is the negated form of $\vec{v}$, exactly like we defined L. As a result, it is possible to represent a linear transformation by adding two scaled vectors:

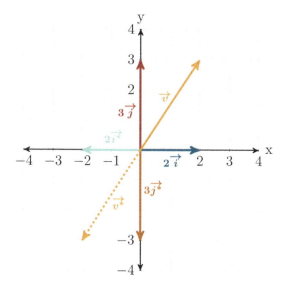

Figure 4.3: The transformation of both $\vec{v}$ and its basis.

But What About a Matrix?

Generically, if $\vec{v} = (v_1, v_2)$ is any vector of $\mathbb{R}^2$, a linear transformation can have the form:

$$\vec{v^*} = v_1 \cdot \vec{i^*} + v_2 \cdot \vec{j^*}$$

To transform a given vector from $\mathbb{R}^2$, we can start by mapping the standard basis. Then, suppose we add the result of multiplying each of the elements of the original vector by the vectors of this transformed version of the basis, respectively. In that case, we will obtain the desired transformed vector. Let's consider another example to solidify this important concept. Say that we now have G, a linear transformation where $\vec{i}$ and $\vec{j}$ are transformed into $(2, 1)^T$ and $(1, 2)^T$. A vector that results from applying G can be described as $\vec{g^*}$:

$$\vec{g^*} = g_1 \cdot \begin{pmatrix} 2 \\ 1 \end{pmatrix} + g_2 \cdot \begin{pmatrix} 1 \\ 2 \end{pmatrix} \tag{4.3}$$

We are missing the vector from which we desire to obtain a transformed version, the vector $\vec{g}$. Let's say that g_1 and g_2, the elements of $\vec{g}$, are equal to 1. Consequently, we have:

$$\vec{g} = \begin{pmatrix} g_1 \\ g_2 \end{pmatrix} = \begin{pmatrix} 1 \\ 1 \end{pmatrix}$$

So, it follows that:

$$\vec{g^*} = 1 \cdot \begin{pmatrix} 2 \\ 1 \end{pmatrix} + 1 \cdot \begin{pmatrix} 1 \\ 2 \end{pmatrix} = \begin{pmatrix} 3 \\ 3 \end{pmatrix}$$

If you are curious, this is the analytic representation of G:

$$G : \mathbb{R}^2 \to \mathbb{R}^2$$

$$(g_1, g_2) \to (2g_1 + g_2, g_1 + 2g_2)$$

And, visually, this is how we can represent this transformation:

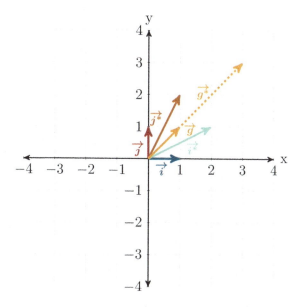

Figure 4.4: The transformation of $\vec{g}$.

Before moving onto a matrix representation of linear transformations, there is something else to cover. We've spoken about changing basis, linear transformations, and changing basis with linear transformations. A change of basis is not a linear transformation. You can do a change of basis via an isomorphism, which is a linear transformation between the same spaces. Remember that changing the basis just changes the perspective of the same vector, whereas a linear transformation, as the name suggests, can change the vector; it is a transformation. For example, consider an arbitrary mapping from $\mathbb{R}^2$ to $\mathbb{R}^3$:

$$H : \mathbb{R}^2 \to \mathbb{R}^3 \tag{4.4}$$

$$(h_1, h_2) \to (7h_1, 2h_2 + h_1, h_1 + h_2)$$

The equation 4.4 can't represent a change of basis, but a linear transformation sure can. It is true that we did not define the conditions that a mapping has to obey in order for it to be considered linear. But it is Sunday, and I was hoping that you would let this one slide... no? Forward we move then. A linear transformation $L : \mathbb{R}^n \to \mathbb{R}^m$ must satisfy two conditions:

1. $L(\vec{v} + \vec{u}) = L(\vec{v}) + L(\vec{u})$

2. $L(c \cdot \vec{v}) = c \cdot L(\vec{v})$

For all vectors $\vec{v}, \vec{u} \in \mathbb{R}^n$ and all scalars $c \in \mathbb{R}$.

Alright, let's verify if H satisfies the two conditions above. On H we go from $\mathbb{R}^2$ to $\mathbb{R}^3$ so we can define $\vec{v}$ and $\vec{u}$ as (v_1, v_2) and (u_1, u_2). The plan of attack will be to get the left side of the item's one equation and develop to see if we arrive at the equation on the right of the same item's number. We need $(\vec{v} + \vec{u})$ which is the same as $(v_1 + u_1, v_2 + u_2)$. Now this specimen will go for a ride via H:

$$H(v_1 + u_1, v_2 + u_2) = (7(v_1 + u_1), 2(v_2 + u_2)$$

$$+(v_1 + u_1), (v_1 + u_1) + (v_2 + u_2))$$

I know that stuff is so ugly that if it could look at a mirror its reflection would walk away, but it is what we have to work with. Now for the second part we need $H(\vec{v})$ and $H(\vec{u})$ to calculate $H(\vec{v}) + H(\vec{u})$. So:

$$H(\vec{v}) = (7v_1, 2v_2 + v_1, v_1 + v_2)$$

On the other hand:

$$H(\vec{u}) = (7u_1, 2u_2 + u_1, u_1 + u_2)$$

Adding them up results in:

$$(7(v_1 + v_2), 2(v_2 + u_2) + v_1 + u_1, v_1 + v_2 + u_1 + u_2)$$

Which is exactly what we are looking for. Now for item number two, let's start with $H(c\vec{v})$:

$$H(cv_1, cv_2) = (7cv_1, 2cv_2 + cv_1, cv_1 + cv_2)$$

$$= (7cv_1, c(2v_2 + v_1), c(v_1 + v_2))$$

$$= c(7v_1, 2v_2 + v_1, v_1 + v_2)$$

$$= cH(v_1, v_2)$$

And that proves point number two! This means that we are assured that H is a linear transformation, and there is the opportunity to learn two more fancy names: for item one we have additivity whereas item two is homogeneity.

I want for us to go back to equation 4.3, where we defined the vector $\vec{g^*}$ via a linear combination:

$$\vec{g^*} = g_1 \cdot \begin{pmatrix} 2 \\ 1 \end{pmatrix} + g_2 \cdot \begin{pmatrix} 1 \\ 2 \end{pmatrix}$$

There is another way of representing such a transformation; we could do so via a matrix:

$$G = \begin{pmatrix} 2 & 1 \\ 1 & 2 \end{pmatrix}$$

To go from one notation to the other, from analytical to a matrix, is very simple. Each row of the matrix will have the coefficients of the variables of each element of the linear equation above. So, the first row of matrix G is $2, 1$ because the coefficients of the first element $(2g_1 + g_2)$ are two and one. The second line of the matrix follows the same logic. The conclusion is that a linear transformation can also be characterised by a matrix.

We now know about vectors and matrices, but we don't have a way of operating between these two concepts, and in fact, this is what we are missing. To transform any vector with that matrix G, we need to multiply a vector by a matrix.

4.2.1 Could It Be Love? Matrix and Vector Multiplication

Matrix and vector multiplication can be summarised by a set of rules. A method that is appealing to some people for reasons that I will leave to your imagination. On the other hand, I have a very hard time with such things, so instead of just providing a formula to compute these multiplications, we will deduce it. We are not cooking; we are learning mathematics. You shouldn't be happy because you have a set of instructions to follow. Save that for your bad-tasting vegan dessert. I am just joking. I know you are all foodies that know more than people that spent half of their lives cooking.

Let's consider any vector, which we'll call $\vec{x}$ (we need a x in a mathematics book, so it was about time!), and a matrix A. There is one consideration to bear in mind when defining matrix-vector products, their dimensions. We can only compute this multiplication when the number of columns in A is equal to the number of rows in $\vec{x}$ or the number of columns in $\vec{x}$ is equal to the number of rows in A. So for any matrix $A_{m \times n}$ we can only compute multiplication with vectors of the order $\vec{x}_{n \times 1}$. We can then represent, $A \cdot \vec{x}$ as:

$$\begin{pmatrix} a_{11} & a_{12} & \cdots & a_{1n} \\ a_{21} & a_{22} & \cdots & a_{2n} \\ \vdots & \vdots & \ddots & \vdots \\ a_{m1} & a_{m2} & \cdots & a_{mn} \end{pmatrix} \cdot \begin{pmatrix} x_1 \\ x_2 \\ \vdots \\ x_n \end{pmatrix} \tag{4.5}$$

Time for a confession. The term role model would not have applied to me as a student. Under my name on the attendance report, things were not pretty. This lack of showing up is not something that I am proud of these days. I could have learned way more in university, but everything has its positives and its negatives. I missed a lot of classes, but I never flunked. This is because I developed a strategy. Where and when I took my degree, it was not mandatory to go to class; you could just show up on the exam date, and if you scored more than 50%, the job was done. You can't fully understand a whole semester of classes a week before the exam. That was an easily reachable conclusion. So now I was seriously debating, how could I pass this exam by not attending classes all the time? In the discipline of mathematics, you tend to encounter a dependence on prior knowledge to understand what will come next. You could see it as a problem, but I chose to see it as an advantage. The things you learned in the first weeks of classes will likely be a basis for more complex deductions, concepts, and theorems. So, if you have a solid foundation but need to get to more complex themes, then with some creativity, inspiration, and maybe a little bit of lady luck on your side, you could... wing it. So my strategy, as time was of the essence when studying for an exam, was to just focus on the basics: everything I needed that allowed me to deduce a concept if required.

I am not advocating for people to do this. The fact that I did not attend classes as often as I should have is the sole reason behind my average grades during my degree. Gladly, this changed during my Master's. I worked harder. The positive from this is that I always ensure that I understand the basics well, which has also helped me immensely in my career as a mathematician that has the job title of data scientist. Now in this book, is there anything that we can use to do something to a big ass matrix and a vector? Well, what about the dot product? Let's try it. If we compute the dot product row by row on equation 4.5, we will end up with:

$$\begin{pmatrix} a_{11}x_1 + a_{12}x_2 + \cdots + a_{1n}x_n \\ a_{21}x_1 + a_{22}x_2 + \cdots + a_{2n}x_n \\ \vdots \\ a_{m1}x_1 + a_{m2}x_2 + \cdots + a_{mn}x_n \end{pmatrix}$$

Let's go back to where we finished the exposition of linear transformations with matrix G:

$$G_{2\times2} = \begin{pmatrix} 2 & 1 \\ 1 & 2 \end{pmatrix}$$

And we need a 2×1 vector, so let's put $\vec{g}$ to work again:

$$\vec{g} = \begin{pmatrix} 1 \\ 1 \end{pmatrix}$$

If we were to multiply G and $\vec{g}$:

$$\begin{pmatrix} 2 & 1 \\ 1 & 2 \end{pmatrix} \cdot \begin{pmatrix} 1 \\ 1 \end{pmatrix} = \begin{pmatrix} 2 \cdot 1 + 1 \cdot 1 \\ 1 \cdot 1 + 2 \cdot 1 \end{pmatrix} = \begin{pmatrix} 3 \\ 3 \end{pmatrix}$$

But What About a Matrix?

Given that we used G and $\vec{g}$ on the example above is not a surprise that if we wish to transform a vector by making use of a linear transformation, we just need to multiply this vector by the matrix. Now, let's define the matrix for the linear transformation H:

$$H = \begin{pmatrix} 7 & 0 \\ 2 & 1 \\ 1 & 1 \end{pmatrix}$$

Alright, let's do some verifications with $\vec{g}$:

$$\begin{pmatrix} 7 & 0 \\ 2 & 1 \\ 1 & 1 \end{pmatrix} \cdot \begin{pmatrix} 1 \\ 1 \end{pmatrix} = (7, 3, 2)$$

The linear transformation H is defined such that the new vector is given by $(7h_1 + 0h_2, 2h_2 + h_1, h_1 + h_2)$ so if we replace h_1 with 1 and h_2 with 1 we end up with the same result $(7, 3, 2)$. Cool stuff! This is another use case for a matrix, the representation of a linear transformation. And on the way to understanding it, we introduced matrix-vector multiplications. If a matrix is a set of vectors and we can multiply a vector by a matrix, what is stopping us from multiplying two matrices? Surely not me. Given two generic matrices $A_{n \times m}$ and $B_{m \times p}$, actually throw one more into the mix, it is happy hour, let's say that $C_{n \times p}$ is the result of $A \cdot B$, so if:

$$A = \begin{pmatrix} a_{11} & a_{12} & \cdots & a_{1m} \\ a_{21} & a_{22} & \cdots & a_{2m} \\ \vdots & \vdots & \ddots & \vdots \\ a_{n1} & a_{n2} & \cdots & a_{nm} \end{pmatrix}$$

And:

$$B = \begin{pmatrix} b_{11} & \cdots & b_{1p} \\ b_{21} & \cdots & b_{2p} \\ \vdots & \ddots & \vdots \\ b_{m1} & \cdots & b_{mp} \end{pmatrix} = \left[\begin{bmatrix} b_{11} \\ b_{21} \\ \vdots \\ b_{m1} \end{bmatrix} \cdots \begin{bmatrix} b_{1p} \\ b_{2p} \\ \vdots \\ b_{mp} \end{bmatrix} \right]$$

Then each column of C is the matrix-vector product of A with the respective column in B. In other words, the component in the i_{th} row and j_{th} column of C is the dot product between the i_{th} row of A and the j_{th} column of B such that:

$$c_{ij} = a_{i1} \cdot b_{1j} + a_{i2} \cdot b_{2j} + \ldots + a_{in} \cdot b_{nj}$$

Visually this is what we will have:

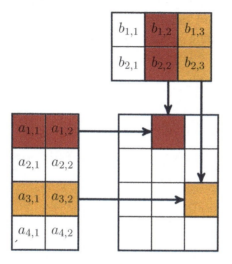

Figure 4.5: Coloured squares to help us with matrix multiplication.

Let's check a numerical example. Say that $A_{4 \times 2}$ is such that:

$$A = \begin{pmatrix} 2 & 4 \\ 1 & 1 \\ 3 & 2 \\ 1 & 1 \end{pmatrix}$$

And let's define $B_{2 \times 3}$:

$$B = \begin{pmatrix} 1 & 2 & 1 \\ 3 & 1 & 2 \end{pmatrix}$$

Then:

$$C = A \cdot B = \begin{pmatrix} 2 & 4 \\ 1 & 1 \\ 3 & 2 \\ 1 & 1 \end{pmatrix} \cdot \begin{pmatrix} 1 & 2 & 1 \\ 3 & 1 & 2 \end{pmatrix}$$

$$= \begin{pmatrix} 2 \cdot 1 + 4 \cdot 3 & 2 \cdot 2 + 4 \cdot 1 & 2 \cdot 1 + 4 \cdot 2 \\ 1 \cdot 1 + 1 \cdot 3 & 1 \cdot 2 + 1 \cdot 1 & 1 \cdot 1 + 1 \cdot 2 \\ 3 \cdot 1 + 2 \cdot 3 & 3 \cdot 2 + 2 \cdot 1 & 3 \cdot 1 + 2 \cdot 2 \\ 1 \cdot 1 + 1 \cdot 3 & 1 \cdot 2 + 1 \cdot 1 & 1 \cdot 1 + 1 \cdot 2 \end{pmatrix} = \begin{pmatrix} 14 & 8 & 10 \\ 4 & 3 & 3 \\ 9 & 8 & 7 \\ 4 & 3 & 3 \end{pmatrix}$$

It is simply not enough. We are manipulating vectors with dot products and dealing with linear transformations. Something is happening in the vector space, and it shouldn't go to the Vatican to be kept a secret. Consider:

$$Z : \mathbb{R}^2 \to \mathbb{R}^2$$

$$(z_1, z_2) \to (-z_1 + z_2, 2z_1 + z_2)$$

The matrix that represents the linear transformation Z is:

$$\begin{pmatrix} -1 & 1 \\ 2 & 1 \end{pmatrix}$$

Transforming the vector $\vec{g}$ with the linear transformation Z results in:

$$\begin{pmatrix} -1 & 1 \\ 2 & 1 \end{pmatrix} \cdot \begin{pmatrix} 1 \\ 1 \end{pmatrix} = (0, 3) = \vec{g_Z^*}$$

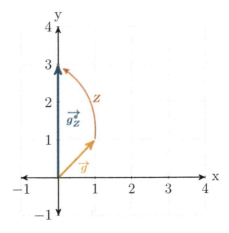

Figure 4.6: A transformation of $\vec{g}$ by Z.

Now I will define another linear transformation. We will get somewhere, don't worry, I haven't lost my mind. I don't know how to convince you of this, as I think that everybody that loses their mind will state that same sentence, nevertheless, give me a chance. Let's say that W is defined such that:

$$W : \mathbb{R}^2 \to \mathbb{R}^2$$

$$(w_1, w_2) \to (w_2, w_1 + w_2)$$

It would be interesting to see what happens to vector $\overrightarrow{g_Z^*}$ if we transform it with W. In other words, we are exploring the idea of composing transformations. We are looking at transforming one vector via a matrix and mapping the resultant element with another matrix.

So, we will transform the vector that results from changing $\overrightarrow{g}$ with Z via the matrix W:

$$\begin{pmatrix} 0 & 1 \\ 1 & 1 \end{pmatrix} \cdot \begin{pmatrix} 0 \\ 3 \end{pmatrix} = (3,3) = \overrightarrow{g_W^*}$$

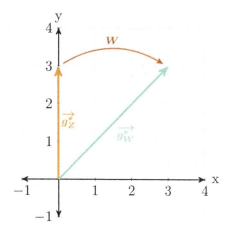

Figure 4.7: A transformation of $\overrightarrow{g_Z^*}$ by W.

But What About a Matrix?

The vector $\overrightarrow{g_W^*}$ is no stranger to us. It is the same as the vector $\overrightarrow{g^*}$ we obtained when we introduced the subject of linear transformations and used the mapping G as an example. This did not happen by accident, I fabricated this example to showcase a composition, so let's multiply the two matrices W and Z:

$$\begin{pmatrix} 0 & 1 \\ 1 & 1 \end{pmatrix} \cdot \begin{pmatrix} -1 & 1 \\ 2 & 1 \end{pmatrix} = \begin{pmatrix} 2 & 1 \\ 1 & 2 \end{pmatrix}$$

This is the matrix that defines the linear transformation G, which means that when we are multiplying matrices, we are composing a transformation. We are transforming a vector in one step instead of two. While two applies to this particular case, you can have as many steps as there are matrices that you wish to multiply. This is a good way of looking at matrix multiplication rather than just thinking of a formula. Now you can see that you are just taking a vector for a ride. Either you do it in steps if you wish to multiply the vector by one of the matrices in the multiplication, and then multiply the result of this by the next matrix and so on. Or you can do it in one go and multiply the vector by the resultant matrix of the multiplication.

Never forget that order matters, oops I forgot to mention it, so let's introduce some matrix multiplication concepts. Contrary to multiplication with real numbers, multiplying two matrices is not commutative, meaning that if A and B are two matrices:

$$A \cdot B \neq B \cdot A$$

The dimensions of the matrices are also critical, you can't just multiply any two matrices, just like the case of matrix-vector multiplication. The number of columns in the first matrix must be equal to the number of rows in the second matrix. The resultant matrix will then have the number of rows of the first matrix and the number of columns of the second matrix:

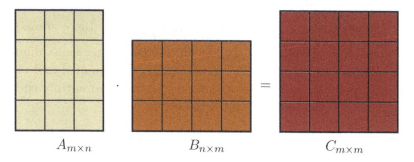

Figure 4.8: Brown, orange and red so the dimensionality isn't a pain in the head.

There are several ways to multiply matrices, but as we are dealing with linear mappings (we will also introduce linear systems later), we need to pay attention to the dimensions. Finally, the product of matrices is both distributive and associative when dealing with matrix addition:

- $A(B + C) = AB + AC$

- $(B + C)D = BD + CD$

I feel that we are on a roll here with these concepts of multiplication, so I will take the opportunity to present a new way of representing the multiplication of two matrices, if I may. Let's consider two generic matrices, A and B:

$$A = \begin{pmatrix} a & b \\ c & d \end{pmatrix} \quad \text{and} \quad B = \begin{pmatrix} e & f \\ g & h \end{pmatrix}$$

If we multiply A and B it follows that:

$$A.B = \begin{pmatrix} ae + bg & af + bh \\ ce + dg & cf + dh \end{pmatrix}$$

We can split this into the sum of two matrices:

$$A.B = \begin{pmatrix} ae & af \\ ce & cf \end{pmatrix} + \begin{pmatrix} bg & bh \\ dg & dh \end{pmatrix} \tag{4.6}$$

61

Right, let's take a look at matrix $\begin{pmatrix} ae & af \\ ce & cf \end{pmatrix}$, there are a's, e's, c's, and f's. Don't worry; we will not be singing songs about the alphabet. Those letters are in the first column of A and the first row of B, and it seems that I just have to multiply each entry of one with each entry of the other:

$$\begin{pmatrix} a \\ c \end{pmatrix}(e \quad f) = \begin{pmatrix} ae & af \\ ce & cf \end{pmatrix}$$

This technique is called the outer product and it is represented by the symbol $\otimes$. For some reason, for a while, I would have an anxiety attack every time I saw that symbol on a blackboard at university, and now I understand why I feared the notation. There is no reason to fear an equation. They are the syntax of mathematics, the same way that music has symbols and shapes which form notes and tempos. So in terms of notation, if we have vector $\vec{v}$ with dimensions $n \times 1$ and vector $\vec{w}$ with dimensions $m \times 1$, and their outer product is computed as:

$$\vec{v} \otimes \vec{w} = \begin{pmatrix} v_1 \cdot w_1 & v_1 \cdot w_2 & \cdots & v_1 \cdot w_n \\ v_2 \cdot w_1 & v_2 \cdot w_2 & \cdots & v_2 \cdot w_n \\ \vdots & \vdots & \ddots & \vdots \\ v_n \cdot w_1 & v_n \cdot w_2 & \cdots & v_n \cdot w_m \end{pmatrix}$$

This is an operation between two vectors. But, the reason I chose to introduce it while we are dealing with matrices is: well, the first reason is that the result is a matrix. As for the second reason, we can put this operation into context. Going back to equation 4.6, we divided the multiplication of A and B into a sum of two matrices, and followed this with a new way of computing a matrix via the outer product of two vectors. This technique allows us to represent matrix multiplication as a sum of outer products. The outer product of the vectors (b, d) and (g, h) provides the right -

side of the summation on equation 4.6, so we can generalise the formula for multiplication of matrices as a sum of dots products:

$$A.B = \sum_{i=1}^{\#columnsA} \mathrm{Column}_i A \otimes \mathrm{Row}_i B$$

Where $\mathrm{Column}_i A$ is the i_{th} column of A and $\mathrm{Row}_i B$ is i_{th} row of B. There are several applications for representing matrix multiplication as the sum of outer products. For example, if you have a sparse matrix (a matrix in which most entries are zero) you can speed up the computation taken to multiply two matrices by this representation. Another example is the derivation of an approximation of matrices via matrix decomposition, such as a single value decomposition, which we will cover in this book, so bear with me.

By looking at linear transformations, we had the need to understand several other concepts that while not complicated, were of extreme importance. We went on a ride there with the linear transformations (just like the vectors), and I can tell you that by now, you are most likely capable of understanding any concept in linear algebra. Just think about what we've covered so far: vectors, dot products, matrices, linear transformations, outer products, the multiplication of vectors by matrices, as well as matrices by matrices. One cool thing you can do with linear transformations is to transform shapes. For example, you can shrink or extend a square into a rectangle, and with this comes a useful concept in linear algebra, the determinant.

4.2.2 When It's Not Rude to Ask About Size - The Determinant

Linear transformations are a fundamental concept in linear algebra; they allow the rotation and scaling of vectors in the same space or even into new ones. We can also use this mathematical concept to transform geometric figures, like a square or a parallelogram. If we take a two-dimensional space, such formations can be represented by a pair of vectors and their corresponding projections. Let's use the standard basis $\overrightarrow{i} = (1,0)^T$ and $\overrightarrow{j} = (0,1)^T$ and form a square:

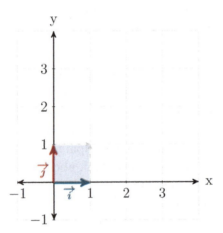

Figure 4.9: First example of the determinant saga.

Following on from this, we can define a linear transformation represented by the matrix:

$$L = \begin{pmatrix} 2 & 0 \\ 0 & 2 \end{pmatrix}$$

Applying the mapping L to $\overrightarrow{i}$ and $\overrightarrow{j}$ will result into two new vectors, $\overrightarrow{i^*}$ and $\overrightarrow{j^*}$, with values $(2,0)^T$ and $(0,2)^T$ respectively. If we plot them alongside their projections, this is what it will look like:

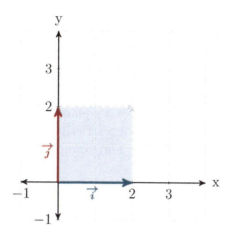

Figure 4.10: Second example of the determinant saga.

Visually, the area of the square formed by the basis vectors quadruples in size when transformed by L. So, every shape that L modifies will be scaled by a factor of four. Now, consider the following linear transformation H defined by:

$$H = \begin{pmatrix} \frac{1}{2} & 0 \\ 0 & \frac{1}{2} \end{pmatrix}$$

If we apply H to the vectors $\vec{i}$ and $\vec{j}$ we will end up with $\overrightarrow{i^{**}}$ and $\overrightarrow{j^{**}}$ with values $(\frac{1}{2}, 0)^T$ and $(0, \frac{1}{2})^T$:

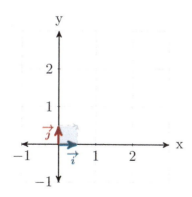

Figure 4.11: Third example of the determinant saga.

In this instance, the area of the parallelogram shrunk by a factor of four. The scalar that represents the change in area size due to the transformation is called the determinant. In the first example, we have a determinant equal to four and the second example has a determinant of one quarter. The determinant is then a function that receives a square matrix and outputs a scalar that does not need to be strictly positive. It can take negative values or even zero. For example, consider the transformation Z such that:

$$Z = \begin{pmatrix} -1 & 0 \\ 0 & 1 \end{pmatrix}$$

To spice things up a little, and I mean it when I say a little bit. Once upon a time in Mexico, I decided to play macho and asked for a very spicy dish. So let's say that I and the Mexican sewer system had some rough next few days. Let's consider two new vectors:

$$\vec{g} = \begin{pmatrix} 3 \\ 2 \end{pmatrix}, \quad \vec{l} = \begin{pmatrix} 1 \\ 2 \end{pmatrix}$$

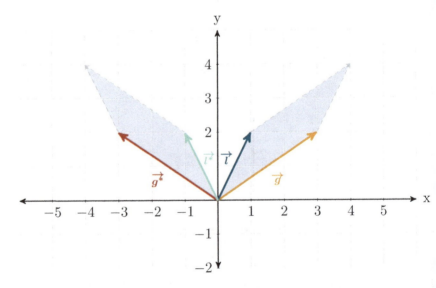

Figure 4.12: Fourth example of the determinant saga.

The image above represents a determinant with a negative value. It might seem strange as the determinant represents a scaling factor. But, the negative sign indicates that the orientation of the vectors has changed. The vector $\overrightarrow{g}$ was to the right of $\overrightarrow{l}$ and after the transformation, $\overrightarrow{g}$ is to the left of $\overrightarrow{l}$. Due to this property, the determinant can also be called the signed area. What we are missing is a formula to compute this bad boy, so let's consider a generic matrix M:

$$M = \begin{pmatrix} a & b \\ c & d \end{pmatrix}$$

The notation for the determinant is $det(M)$ or $\mid M \mid$, and the formula for its computation is:

$$det \begin{pmatrix} a & b \\ c & d \end{pmatrix} = \begin{vmatrix} a & b \\ c & d \end{vmatrix} = ad - bc \tag{4.7}$$

Why? It is a fair question. Let's bring up a plot to understand what is going on. We have been speaking about areas, so a visual should indicate why the formula is as shown above:

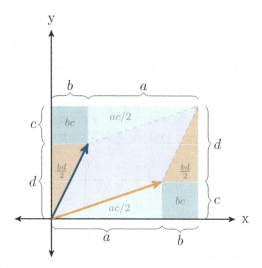

Figure 4.13: The geometry of a 2×2 determinant.

The quest now is for the area of the blue parallelogram. At first glance, it does not seem like a straightforward thing to compute, but we can find a workaround. We can start by calculating the area of the big rectangle, formed by $(a + b)$ and $(c + d)$, this can be found by $(a + b)(c + d)$. We don't know the area of the blue parallelogram, but we know that of the shapes around it: four triangles and two squares. So, if we subtract the sum of these from the big rectangle, we end up with the area value of the blue parallelogram. The two brown triangles each have an area of $\frac{bd}{2}$, which when summed is bd, so we need to remove bd from the big parallelogram. The two pink squares are each of size bc, which when summed equates to a value of $2bc$. And finally, the green triangles are each of size $\frac{ac}{2}$, adding them will result in a magnitude of ac. If we sum each of these areas, we arrive at an expression for the value that we need to subtract from the big rectangle area:

$$\begin{vmatrix} a & b \\ c & d \end{vmatrix} = (a + b)(c + d) - ac - bd - 2bc = ad - bc$$

Alright, let's check what happens if we apply equation 4.7 to the linear transformation Z:

$$\begin{vmatrix} -1 & 0 \\ 0 & 1 \end{vmatrix} = -1 \cdot 1 - 0 \cdot 0 = -1$$

This means that the resultant area of a shape transformed by Z will remain the same, but the orientation of the vectors will change, just as we saw in the plot above. If we have a higher dimensionality than two, we will be scaling volumes instead of scaling areas. Consider a generic 3×3 matrix, N such that:

$$N = \begin{pmatrix} a & b & c \\ d & e & f \\ g & h & i \end{pmatrix}$$

Graphically, the determinant will reflect the change in volume of a transformed parallelepiped:

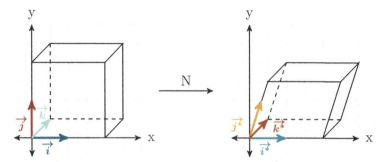

Figure 4.14: The geometry of a 3 × 3 determinant.

Before jumping into the derivation of the formula to compute a 3 × 3 determinant, I would like to introduce three properties of this algebraic concept:

- If we have pair of vectors that are equal, its follows that $det(\vec{i}, \vec{j}, \vec{k}) = 0$. Graphically we can prove this. For example, if we collapse $\vec{j}$ into $\vec{i}$, we understand that the parallelepiped will also collapse and become flat.

- If we scale the length of any side of the parallelepiped, we will also scale the determinant, $det(a \cdot \vec{i}, \vec{j}, \vec{k}) = a \cdot det(\vec{i}, \vec{j}, \vec{k})$, where $a \in \mathbb{R}$.

- The determinant is a linear operation, $det(\vec{i} + \vec{w}, \vec{j}, \vec{k}) = det(\vec{i}, \vec{j}, \vec{k}) + det(\vec{w}, \vec{j}, \vec{k})$.

These properties are valid for any dimensionality of a geometric shape that we wish to calculate a determinant for, but for now, I would like to do an experiment and use one of these properties to produce the same formula that we have for a matrix of size 2 × 2. For this, let's formulate the calculation of a determinant such that:

$$det(M) = det\left(\begin{pmatrix} a \\ c \end{pmatrix}, \begin{pmatrix} b \\ d \end{pmatrix} \right)$$

We can represent these vectors as a linear combination of the standard basis vectors $\vec{i}$, $\vec{j}$:

$$det(a \cdot \vec{i} + c \cdot \vec{j}, b \cdot \vec{i} + d \cdot \vec{j}) \qquad (4.8)$$

I feel that there are a ton of letters and no numbers, and I don't want to scare you off, so let's do a verification:

$$a \cdot \begin{pmatrix} 1 \\ 0 \end{pmatrix} + c \cdot \begin{pmatrix} 0 \\ 1 \end{pmatrix} = \begin{pmatrix} a \\ c \end{pmatrix}$$

There! Some numbers to ease us into what's about to happen. We can make equation 4.8 look more friendly by leveraging property numero tres:

$$det(a \cdot \vec{i}, b \cdot \vec{i} + d \cdot \vec{j}) + det(c \cdot \vec{j}, b \cdot \vec{i} + d \cdot \vec{j})$$

This can also be made a bit simpler:

$$det(a \cdot \vec{i}, b \cdot \vec{j}) + det(a \cdot \vec{i}, d \cdot \vec{j})$$

$$+ det(c \cdot \vec{j}, b \cdot \vec{i}) + det(c \cdot \vec{j}, d \cdot \vec{i})$$

Now, by property two we can take those scalars out:

$$ab \cdot det(\vec{i}, \vec{i}) + ad \cdot det(\vec{i}, \vec{j}) \qquad (4.9)$$

$$+ cb \cdot det(\vec{j}, \vec{i}) + cd \cdot det(\vec{j}, \vec{j})$$

OK, we can do something with this. If a matrix has identical vectors for columns, it follows that the determinant value is 0. So, $ab \cdot det(\vec{i}, \vec{i}) = 0$ and $cd \cdot det(\vec{j}, \vec{j}) = 0$. The only two things that are missing are the values of $det(\vec{i}, \vec{j})$ and $det(\vec{j}, \vec{i})$. This will be the same matrix, but with switched rows. Let's see what happens to the determinant when we do such a thing:

$$\begin{vmatrix} 1 & 0 \\ 0 & 1 \end{vmatrix} = 1 \cdot 1 - 0 \cdot 0 = 1$$

And:

$$\begin{vmatrix} 0 & 1 \\ 1 & 0 \end{vmatrix} = 0 \cdot 0 - 1 \cdot 1 = -1$$

So, equation 4.9 is equal to:

$$ad - cb$$

Which is what we found when we made the geometric deduction. I included this because we can do the same kind of exercise for a matrix with dimension 3×3, for example, the matrix N. This time I will save you from all the intermediate steps (as they are very similar to those for two dimensions) and I'll provide you with the final result:

$$det(N) = aei + bfg + cdh - ceg - bdi - afh \qquad (4.10)$$

We can reorganise the equation 4.10 to:

$$det(N) = a(ei - fh) - b(di - fg) + c(dh - eg) \qquad (4.11)$$

And we can also produce a less condensed version of equation 4.11

$$det(N) = a\begin{vmatrix} e & f \\ h & i \end{vmatrix} - b\begin{vmatrix} d & f \\ g & i \end{vmatrix} + c\begin{vmatrix} d & e \\ g & h \end{vmatrix} \qquad (4.12)$$

Which has the following graphical representation:

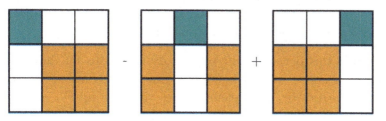

Figure 4.15: Green and yellow for the calculations to be mellow.

For any matrix with a dimensionality lower than four, we are sorted. The only matrix sizes we are missing are four, five, six, and seven ... damn, there is still a lot to cover! It is formula time. For that, let's introduce a new bit of notation. I chose not to do this earlier as I felt it would have become a bit intense too soon, but for what we have to understand now, it will be of extreme utility. Let's redefine the entries of N as:

$$N = \begin{pmatrix} n_{11} & n_{12} & n_{13} \\ n_{21} & n_{22} & n_{23} \\ n_{31} & n_{32} & n_{33} \end{pmatrix}$$

So if we write equation 4.10 following this new notation, it comes that:

$$det(N) = n_{11} \cdot n_{22} \cdot n_{33} + n_{12} \cdot n_{23} \cdot n_{31} + n_{13} \cdot n_{21} \cdot n_{32} \quad (4.13)$$

$$-n_{13} \cdot n_{22} \cdot n_{31} - n_{12} \cdot n_{21} \cdot n_{33} - n_{11} \cdot n_{23} \cdot n_{32}$$

As before, we have to work with what we have. Equation 4.13 is made up of sums and subtractions of the different iterations of multiplications of the elements of N (that's a mouthful, I know!). The suggestion is that we could probably use the Greek letter Σ (used for summations) to represent what we are trying to deduce, that being a general formula for the determinant. It sounds like a good idea, but if we add elements, how will we take care of those negative signs in equation 4.13? We have six terms when calculating the determinant of a three-dimensional matrix, and two when working with a matrix in two dimensions. It follows that the number of terms in the determinant equation is equal to the factorial of the matrix dimensions.

Skip this paragraph if you already know this bit, but a factorial is represented by !, and it refers to the combinations one can construct by using elements described by n. For example, in the case that $n = 3$, we have six ways of combining three components, as reflected in equation 4.13. The formula for the factorial is then the multiplication of each positive integer less than or equal to n, so $3! = 3 \cdot 2 \cdot 1 = 6$.

Alright, so I now know the number of elements for the determinant of a given matrix size, but how will this help with the plus and minus signs? This is where we have to introduce the notion of a permutation, which is a way to order a set of numbers, and in the case of three dimensions, we have six different permutations.

For example consider the numbers $(1, 2, 3)$: we can have six forms of representing this set $(1, 2, 3), (1, 3, 2), (2, 1, 3), (2, 3, 1), (3, 1, 2)$ and $(3, 2, 1)$. We can define a permutation function π that represents these iterations, for example:

$$\pi : \{1, 2, 3\} \rightarrow \{3, 2, 1\} \tag{4.14}$$

Equation 4.14 represents the changes to the original set in order to produce the permutated set. One goes to three, two stays the same, and three goes to one. So $\pi(1) = 3$. These functions come with a definition that we can leverage. It is called the parity of a permutation, and it tells us the sign of such a mathematical concept. Specifically, it states that the number of inversions defines the sign of a permutation. So, it will be $+1$ if we have an even number of inversions and -1 if the value is odd.

Curiosity has struck me more than once on why non-even numbers are called odd; if you want my honest answer after spending many hours researching this, my conclusion is that I wish I could have that time back! On my journey, I went all the way from shapes defined by the ancient Greeks, to bathroom philosophy on world views and representations, all to no avail.

Anyhow to calculate the number of inversions, we need to check it for pairs. It then follows that we will have an inversion for a pair (x, y), if we have $x < y$, and $\pi(x) > \pi(y)$. Let's go back to the example in equation 4.14, but this time for the pairs when $x < y$: $(1, 2), (1, 3)$ and $(2, 3)$. Now $\pi(1) = 3$, $\pi(2) = 2$ and $\pi(3) = 1$ so:

- $1 < 2$ and $\pi(1) > \pi(2)$ it is an inversion.

- $1 < 3$ and $\pi(1) > \pi(3)$ it is an inversion.

- $2 < 3$ and $\pi(2) > \pi(3)$ it is an inversion.

We arrive at a result where the sign of $(3, 2, 1)$ is negative. If we look into the monomials of equation 4.13 we could define permutations and, consequently, functions from the indices of the elements that are part of them. If you are wondering, 4.13 is a polynomial made up of monomials. Monomial is just a lame name for product of variables and/or constants, where each variable has a non-negative integer exponent. A polynomial is a set of monomials, which make it super lame? I leave this one for you to decide. For example $n_{11}.n_{22}.n_{33}$ is one of these. Since we are speaking about this particular guy, let's try to understand what exactly I'm talking about. We have three pairs $(1,1)$, $(2,2)$, and $(3,3)$, which means that the function π that represents this is:

$$\pi : \{1,2,3\} \to \{1,2,3\}$$

This case is just like an election; nothing changes! So, the number of inversions is equal to zero, and therefore the sign is positive. This means that if we do this exercise for the five remaining monomials and if the parity sign is the same as that in the equation, we may have found what we were looking for: a way to control the minus or plus sign in a formula to calculate a determinant via a summation. Let's check:

$$
det(N) = \underbrace{\overset{\pi:123}{n_{11}.n_{22}.n_{33}}}_{\text{even}\to\text{sign is }+} - \underbrace{\overset{\pi:132}{n_{11}.n_{23}.n_{32}}}_{\text{odd}\to\text{sign is }-}
$$

$$
- \underbrace{\overset{\pi:213}{n_{12}.n_{21}.n_{33}}}_{\text{even}\to\text{sign is }-} + \underbrace{\overset{\pi:231}{n_{12}.n_{23}.n_{31}}}_{\text{even}\to\text{sign is }+}
$$

$$
+ \underbrace{\overset{\pi:312}{n_{13}.n_{21}.n_{32}}}_{\text{even}\to\text{sign is }+} - \underbrace{\overset{\pi:321}{n_{13}.n_{22}.n_{31}}}_{\text{even}\to\text{sign is }-}
$$

(4.15)

A quick way to detect inversions is to check which pairs appear in the wrong order, for example in the function $\pi : 312$, the pairs $(1,3)$ and $(2,3)$ are inverted; therefore, the sign must be positive. It is true that with just this, we can create a formula for the calculation of the determinant. Now, it will be ugly as hell, but it will work:

$$
det(A) = \sum_{\pi \text{ permutation of } n} sign(\pi)a_{1\pi(1)} \cdot a_{2\pi(2)}...a_{n\pi(n)} \quad (4.16)
$$

A is a square matrix and n is the total number of permutations. We can test this formula with a 2×2 matrix M. So, in the case of a matrix of size two, we have the same number of permutations as for the size, $2! = 2.1 = 2$. They will be $(1,2)$ and $(2,1)$, so we will sum two terms. The first is related to the permutation $(1,2)$, and we need to define π for this case: $\pi : \{1,2\} \rightarrow \{1,2\}$. It follows that $\pi(1) = 1$ and $\pi(2) = 2$, so the first term of the equation is $sign(\pi) \cdot m_{11} \cdot m_{22}$. The $sign(\pi)$ is positive because nothing changes: there is no inversion. If we apply the same logic to the second term, we will have $sign(\pi) \cdot m_{12} \cdot m_{21}$ as $\pi(1) = 2$ and $\pi(1) = 2$. In this case, we have one inversion, so the sign of π is negative. It then follows that:

$$det(M) = m_{11} \cdot m_{22} - m_{12} \cdot m_{21}$$

Which is exactly the same as equation 4.7. It is true that equation 4.16 is enough to calculate any determinant, but we can do better than that. I mean, come on, that will be a nightmare for dimensions higher than three! Something curious happens in equation 4.12; we have three pivots and three lower dimensioned determinants. Let's start with the determinants to write equation 4.12 as:

$$det(N) = a \cdot M_{11} - b \cdot M_{12} + c \cdot M_{13} \qquad (4.17)$$

There are new individuals in equation 4.17, the determinants represented by the letter M. Allow me to introduce you to the minors. A minor is the determinant of a square matrix that results from removing some rows and columns from the original matrix. The clue on which row and column to remove is in its index, so, for example, M_{11} suggests that we should remove the first row and the first column from the original matrix:

$$M_{11} = \begin{vmatrix} a & b & c \\ d & e & f \\ g & h & i \end{vmatrix} = \begin{vmatrix} e & f \\ h & i \end{vmatrix}$$

The only elements we haven't spoken about from equation 4.12 are the pivots, another name I know, but these are just the fellows a, b and c. They happen to all be on the same row. We can generate a new formula for the determinant with all of this information. I have a suggestion, what if we fix a row or a column? Let's start with a column, and we can define the determinant such that:

$$det(A) = \sum_{j=1}^{3}(-1)^{i+j}a_{ij}M_{ij} \tag{4.18}$$

If we set $i = 1$, we obtain equation 4.17. The question now, is whether equation 4.18 works for n dimensions? And the answer is yes, it does. But, there's more we can say, as this theorem has a special name, the Laplace expansion:

$$det(A) = \sum_{j=1}^{n}(-1)^{i+j}a_{ij}M_{ij}$$

The Laplace expansion works for any row with a fixed column, or for any column if we fix the row. The determinant formula has been broken down, but there is another linear algebra concept we can learn from this, the cofactors. Don't worry. You have enough knowledge to compute these as they are a part of the formula for the determinant. So, if you multiply $(-1)^{i+j}$ by the minor M_{ij}, you will then be computing the cofactor C_{ij}. The result of this will be a real number because we are multiplying the resultant of a determinant by 1 or -1. In the example above, where we derived the minor M_{11}, the cofactor C_{11} is equal to:

$$C_{11} = (-1)^{1+1} \cdot M_{11} = (-1)^{2} \cdot \begin{vmatrix} e & f \\ h & i \end{vmatrix}$$

Meaning that each entry of a square matrix will have a cofactor, and therefore we can define a matrix of cofactors as:

$$C_{n\times n} = \begin{pmatrix} C_{11} & C_{21} & \cdots & C_{n1} \\ C_{12} & C_{22} & \cdots & C_{n2} \\ \vdots & \vdots & \ddots & \vdots \\ C_{1n} & C_{1n} & \cdots & C_{nn} \end{pmatrix}$$

Ending this with an example will be the best way I feel. So, say that we have G, a 3×3 matrix defined as:

$$G = \begin{pmatrix} 1 & 4 & 5 \\ 9 & 2 & 4 \\ 3 & 5 & 10 \end{pmatrix}$$

Let's use the Laplace expansion to compute the determinant. I will select the first row, so $j = 1$. For my pivots I will have positions g_{11}, g_{21} and g_{31} and the signals are $+,-,+$ respectively. For minors with $j = 1$ we have M_{11}, M_{21} and M_{31}. Okay, we have everything we need for this calculation, so let's get it done:

$$det(G) = g_{11} \cdot M_{11} - g_{21} \cdot M_{21} + g_{31} \cdot M_{31}$$

Which is the same as:

$$det(G) = 1 \cdot \begin{vmatrix} 2 & 4 \\ 5 & 10 \end{vmatrix} - 4 \cdot \begin{vmatrix} 9 & 4 \\ 3 & 10 \end{vmatrix} + 5 \cdot \begin{vmatrix} 9 & 2 \\ 3 & 15 \end{vmatrix}$$

The result is
$$det(G) = -117$$

While I don't imagine that anyone will be doing a lot of manual computation of the determinant, understanding it is crucial. And if you think that the only use for it is to understand the scale and direction of transformation, you have got another thing coming, as determinants can also be used to invert matrices.

4.2.3 Let's Go The Other Way Around - Inverse of a Matrix

Would you send a friend from your house to another location without telling this person how to get back? Then why the hell would you do this to a vector? We now know that matrices can represent linear transformations and we also know that this mathematical concept transforms vectors in space. If A is a linear transformation, then, its inverse is also a linear transformation. The difference is that the inverse of A returns the vector back to its initial state. Just as any real number has an inverse, a matrix will also have an inverse, well... OK, I admit not all of them do. In reality, there is a condition that the determinant of such a matrix must meet. If the inverse matrix depends on the determinant, then this same matrix must consequently be a square matrix. There is a particular case for a determinant value that is worth taking a closer look at, namely when the value is

0. If the determinant is a scaling factor, what will it mean if we change the shape by a value of 0? It means that we have lost at least one of the dimensions. In the case of a parallelogram, the result will be a line or a point. An example would be:

$$G = \begin{pmatrix} 1 & 1 \\ 1 & 1 \end{pmatrix} \quad \text{with} \quad det(G) = 1 \cdot 1 - 1 \cdot 1 = 0$$

Consider two vectors $\vec{g}$ and $\vec{t}$ such that:

$$\vec{g} = \begin{pmatrix} 3 \\ 1 \end{pmatrix} \quad \text{and} \quad \vec{t} = \begin{pmatrix} 1 \\ 2 \end{pmatrix}$$

Geometrically, their representation is:

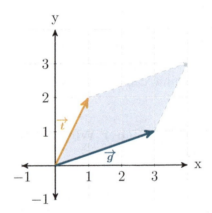

Figure 4.16: The representation of two vectors to aid with the inverse concept

Now let's calculate the coordinates of the new vectors $\vec{g^*}$ and $\vec{t^*}$ which are the vectors that result from transforming $\vec{g}$ and $\vec{t}$ by G. This means that $\vec{g^*} = (4,4)^T$ and $\vec{t^*} = (3,3)^T$:

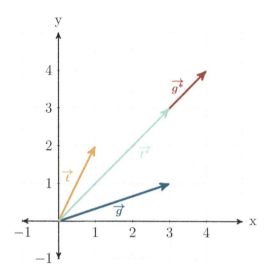

Figure 4.17: Showcase of the mappings of $\overrightarrow{g}$ and $\overrightarrow{t}$ via G.

As expected, the two transformed vectors landed on the same line, and therefore the area of the blue transformed parallelogram is 0.

There is no way to return to the original parallelogram once we arrive on a point or a line, at least with linear transformations. Hence, the second condition for a matrix to have an inverse is that the determinant must not be 0. The last condition is similar to what with real numbers. Consider the inverse of 5. It is defined by $x \in \mathbb{R}$ such that $5.x = 1$, meaning that $x = \frac{1}{5}$. The concept is the same in the case of matrices, but multiplying two matrices won't result in a scalar. It will result in a special matrix called the identity, represented by I_n. The identity is a matrix in which all elements on the diagonal have a value of 1 and all of the non diagonal elements are 0. The matrix diagonal is formed by all the elements $a_{ij} = 1$ where $i = j$:

$$A = \begin{pmatrix} a_{11} & a_{12} & \cdots & a_{1n} \\ a_{21} & a_{22} & \cdots & a_{2n} \\ \vdots & \vdots & \ddots & \vdots \\ a_{n1} & a_{n2} & \cdots & a_{nn} \end{pmatrix}$$

But What About a Matrix?

The identity matrix can then be defined as:

$$I = \begin{pmatrix} 1 & 0 & \cdots & 0 \\ 0 & 1 & \cdots & 0 \\ 0 & 0 & \ddots & 0 \\ 0 & 0 & \cdots & 1 \end{pmatrix}$$

To arrive at a glorious finish with the hope of prosperity, we still need two more things: a visual way to compute the inverse of a matrix, and a numerical example. Moving on to the visualisations:

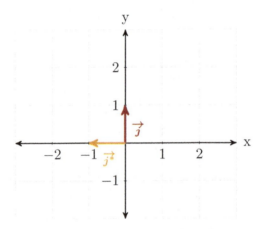

Figure 4.18: A rotation of $\vec{j}$.

In the plot, $\vec{j}$ is $(0,1)^T$ and its transformation $\vec{j^*}$ takes values $(-1,0)^T$. So, the linear transformation responsible for this (responsibility is terrifying when it's said out loud, every time somebody asks 'who's responsible' for something, you know some stuff is coming) can be defined as:

$$M = \begin{pmatrix} 0 & -1 \\ 1 & 0 \end{pmatrix}$$

What we wish to do now is derive M^{-1} such that $M.M^{-1} = I$, right? Yes, this is what we defined previously. As this is a simple example, we know that, in order to calculate M^{-1} we need to define a linear transformation that also does a 90 degree rotation, but in the other direction. So, let's do it in two steps. Firstly, to rotate a vector 90 degrees clockwise, one needs to place the original y coordinate of the vector in the place of the x coordinate and then place the inverse of the initial x coordinate, $-x$, in the place of the original y coordinate, such that:

$$M^{-1} : \mathbb{R}^2 \to \mathbb{R}^2$$

$$(m_1, m_2) \to (m_2, -m_1)$$

Let's test this out. If we transform $\vec{j^*}$ with the linear transformation M^{-1}, we should get the vector $\vec{j}$ such that:

$$\begin{pmatrix} 0 & 1 \\ -1 & 0 \end{pmatrix} \cdot \begin{pmatrix} -1 \\ 0 \end{pmatrix} = \begin{pmatrix} 0 \\ 1 \end{pmatrix}$$

Which is precisely the vector $\vec{j}$. So, in theory, if we multiply M by M^{-1}, we need to get I, the identity matrix. Otherwise, something is not right:

$$M \cdot M^{-1} = \begin{pmatrix} 0 & -1 \\ 1 & 0 \end{pmatrix} \cdot \begin{pmatrix} 0 & 1 \\ -1 & 0 \end{pmatrix}$$

$$= \begin{pmatrix} 0 \cdot 0 + (-1)(-1) & 0 \cdot 1 + 0 \cdot 0 \\ 0 \cdot 1 + (-1) \cdot 0 & 1 \cdot 1 + 0 \cdot 0 \end{pmatrix} = \begin{pmatrix} 1 & 0 \\ 0 & 1 \end{pmatrix}$$

So far, we have defined a few properties regarding the inverse of a matrix, and I feel that it is a good moment to put them in a little box to produce a summary of where we stand. So, if A is such that A^{-1} exists, this means that:

- A is a square matrix.

- A^{-1} is such that $A \cdot A^{-1} = I$.

- The determinant of A is not 0. I will throw another name your way, generosity is the word of the day. When a matrix has this property, it is called non-singular.

The question now is how can we compute an inverse of a matrix? There is a relationship between the determinant of the multiplication of two matrices. We can show that this is equal to the multiplication of the matrix determinants:

$$det(A \cdot B) = det(A) \cdot det(B) \qquad (4.19)$$

If we replace B with the inverse of A:

$$det(A \cdot A^{-1}) = det(I) = 1 \qquad (4.20)$$

From equation 4.19 we can write 4.20 as:

$$det(A) \cdot det(A^{-1}) = 1$$

It follows then that:

$$det(A^{-1}) = \frac{1}{det(A)}$$

Finally we have:

$$A^{-1} = \frac{1}{det(A)}(adj(A))$$

There is some new information in this formula, I am talking about $adj(A)$. Thankfully, we have enough knowledge to derive this very easily. The adjugate is simply the transpose of the cofactors matrix, C, so we can express this as:

$$adj(A) = C^T$$

If you do not recognize the T exponent associated with a matrix, don't worry, papy is here once more to hold your hand. This means that we will swap the rows with the columns, implying that a matrix with dimensions $n \times m$ will become a matrix with dimensions $m \times n$, for example:

$$A = \begin{pmatrix} 1 & 4 \\ 9 & 5 \\ 8 & 2 \end{pmatrix}$$

The transpose of this matrix will still be a matrix, and the notation for it is A^T. The resultant matrix will be:

$$A^T = \begin{pmatrix} 1 & 9 & 8 \\ 4 & 5 & 2 \end{pmatrix}$$

Given that we have computed the transpose of a matrix for a particular case, let's derive the general formula, so we can apply it to any matrix or vector of any dimension. A transposed matrix of dimensions $n \times m$ is a new matrix of size $m \times n$ in which the element in the i_{th} row and j_{th} column becomes the element in the i_{th} column and j_{th} row:

$$\left[A^T\right]_{ij} = [A]_{ji}$$

Don't worry, I did not forget the numerical example! Let's say that the matrix A is defined as:

$$A = \begin{pmatrix} 1 & 1 \\ 0 & 1 \end{pmatrix}$$

Now we need to compute the determinant and the adjugate:

$$det(A) = \begin{vmatrix} 1 & 1 \\ 0 & 1 \end{vmatrix} = 1 \quad \text{and} \quad adj(A) = \begin{pmatrix} 1 & -1 \\ 0 & 1 \end{pmatrix}$$

$$A^{-1} = \frac{1}{1} \cdot \begin{pmatrix} 1 & -1 \\ 0 & 1 \end{pmatrix}$$

Finally it follows that that the inverse of A is:

$$A^{-1} = \begin{pmatrix} 1 & -1 \\ 0 & 1 \end{pmatrix}$$

The matrix A is known to us. We used it when speaking about a change of basis. The vector that we used at the time was $\overrightarrow{v} = (2,3)^T$ and the coordinates of $\overrightarrow{v}$ in the new basis were $(5,3)^T$. So, we must obtain $(2,3)^T$ if we multiply $(5,3)^T$ by A^{-1}. We will now be sending this vector back:

$$\begin{pmatrix} 1 & -1 \\ 0 & 1 \end{pmatrix} \cdot \begin{pmatrix} 5 \\ 3 \end{pmatrix} = \begin{pmatrix} 1 \cdot 5 - 1 \cdot 3 \\ 0 \cdot 5 + 3 \cdot 1 \end{pmatrix} = \begin{pmatrix} 2 \\ 3 \end{pmatrix}$$

Bullseye! That is precisely the result we were looking for, so 50 points for us!

This all started with a linear transformation, but we ended up covering a lot of linear algebra concepts that are needed if we are to fully understand these functions. Mastering mappings that promote movement amongst vectors was crucial in my machine learning journey as a data scientist. The graphical component of such a concept allowed me to think about matrices differently, as a vehicle for movement in space. By doing this, ideas like multiplication of a vector by a matrix, matrix to matrix multiplication, determinants, and inverting matrices became explicit, as I had developed a visual concept of what was happening with these notions when they are applied. If you are wondering where to use all of this mathematics in machine learning, well, neural networks are an excellent example to study. And there is yet another case for matrix representation that is worth exploring: systems of linear equations.

4.3 Gatherings in Space - Systems of Linear Equations

First, let's define a system; and to do that, we will use a vernacular that would impress any executive in an interview room. A system is a collection of components that accepts an input and produces an output. When these systems are linear, we have robust mathematical definitions and concepts that allow us to extract such outcomes, which we can also call solutions. Well, as this book is about linear algebra, this robustness will be provided through methods that we can derive with all of the knowledge we've acquired so far. For the components we have linear equations, and for the inputs and outputs we have vectors. A linear system will then be a set of linear equations that receives a vector and produces one, several, or no solutions. For illustration purposes let's consider a two dimensional scenario as a generic system that we can represent with the following notation:

$$\begin{cases} ax_1 + bx_2 = c \\ dx_1 + ex_2 = f \end{cases} \tag{4.21}$$

In the system 4.21, elements a, b, c, d, e, f are scalars that belong to $\mathbb{R}$ and define the system by characterizing two equations which are lines. The solution is defined by x_1 and x_2, which can be unique, nonexistent, or non-unique.

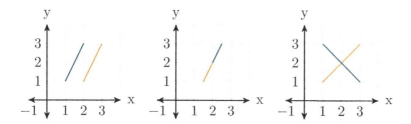

Figure 4.19: The three possible cases for the type of solutions of a system.

The plot above provides a graphical interpretation of some possible type of solutions of a linear system. The blue and yellow lines have for features the scalars a, b, c, d, e, f. If the values are such that the lines are parallel (as shown in the left-hand graph), then the system has no solutions. Conversely, if the lines land on top of each other (as shown in the middle plot), then the system has an infinite number of solutions. Lastly, as seen in the right-hand graph, if the lines intersect each other, this reflects the case where a system has a unique solution. Where are the matrices and the vectors? I miss them too, just like a dolphin misses SeaWorld. But don't worry, we can represent the system 4.21 with matrix-vector notation:

$$A \cdot \vec{x} = \vec{b}$$

The letter A represents a matrix whose elements are the scalars a, b, c, d. The vector $\vec{x}$ is the solution we are looking for (if it exists). Finally, $\vec{b}$ is another vector. The best way to define this vector is to look at linear transformations. If A represents a linear transformation, then what we are saying here is that we want to find a vector $\overrightarrow{x}$, that when transformed by A, lands on $\overrightarrow{b}$. So, $\overrightarrow{b}$ is where the solution has to land after being transformed by A. Our system can be defined by:

$$\begin{pmatrix} a & b \\ d & e \end{pmatrix} \cdot \begin{pmatrix} x_1 \\ x_2 \end{pmatrix} = \begin{pmatrix} c \\ f \end{pmatrix}$$

If we perform that multiplication, we will end up with the same system of equations 4.21, and the symbol { means that we are looking for an interception. However, it is also true that these systems can have as many equations as we wish with as many variables as desired. Consequently, finding a solution for some of them can be portrayed as complicated. Spending a long time doing calculations, only to conclude that there is no solution, is not ideal. Gladly with a matrix representation of such systems, we can understand what is the deal with these so-called solutions, before, doing any calculations towards finding them. We know that for two vectors to intercept somewhere in space, they need to be linearly independent. Therefore, the determinant of a matrix that has these same vectors for columns cannot be zero. From this, it follows that for a system to have a unique solution, the determinant of the matrix that represents the system must be non-zero.

Less talking, more examples, right? Consider a system defined as:

$$\begin{cases} -5x_1 + 2x_2 = -2 \\ -4x_1 + x_2 = -4 \end{cases} \tag{4.22}$$

We now should verify if there is a solution for the system 4.22. For that, we need a matrix A:

$$A = \begin{pmatrix} -5 & 2 \\ -4 & 1 \end{pmatrix}$$

The determinant of A is then calculated by:

$$\begin{vmatrix} -5 & 2 \\ -4 & 1 \end{vmatrix} = -5 \cdot 1 - 2 \cdot (-4) = -5 + 8 = 3$$

The coast seems to be clear. We have a non-zero determinant, meaning that this system will have a unique solution, and we can proceed to learn how to calculate it. But, before jumping into the deduction of a methodology to find solutions for a system, let's quickly go over an example where the determinant is zero. Say that B is:

$$B = \begin{pmatrix} 3 & 1 \\ 6 & 2 \end{pmatrix}$$

The determinant of B is equal to zero, $det(B) = 3 \cdot 2 - 6 \cdot 1 = 0$, and visually:

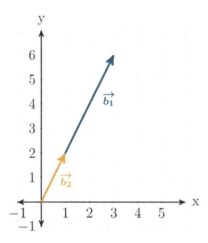

Figure 4.20: Linearly dependent vectors due to a zero value determinant.

As expected, those vectors are linearly dependent. The angle formed by them is 0 degrees. Okay, but the system 4.22 has to have a solution because the determinant of A is not 0. To find it, let's get the vector $\vec{x}$ represented by:

$$\vec{x} = \begin{pmatrix} x_1 \\ x_2 \end{pmatrix}$$

The only thing missing is the vector $\vec{b}$. Now, $\vec{b}$ is not as badass as $\vec{x}$ because, $\vec{x}$ can take any value for coordinates. On the other hand, $\vec{b}$ has fewer options, much like a politician:

$$\vec{b} = \begin{pmatrix} -2 \\ -4 \end{pmatrix}$$

But What About a Matrix?

The equation below represents the system in question:

$$\overbrace{\begin{pmatrix} -5 & 2 \\ -4 & 1 \end{pmatrix}}^{A} \cdot \overbrace{\begin{pmatrix} x_1 \\ x_2 \end{pmatrix}}^{\vec{x}} = \overbrace{\begin{pmatrix} -2 \\ -4 \end{pmatrix}}^{\vec{b}}$$

Let's pause for a moment to observe the representation above, that is if you don't mind. A is a matrix that represents a linear transformation, and $\vec{x}$ is a vector. We know that $A \cdot \vec{x}$ will map $\vec{x}$ into a new vector. The restriction now is that the resultant recast vector of that transformation has to be equal to another vector, $\vec{b}$:

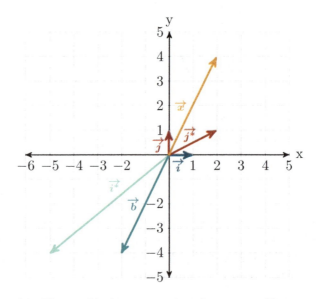

Figure 4.21: The graphical representation of a system of linear equations.

The plot is populated, but let's take it step-by-step. The vectors $\vec{i}$ and $\vec{j}$ represent the standard basis. A generic vector is represented by $\vec{x}$, so please ignore the coordinate $(2, 4)$. It is just for a visual representation. Now, $\vec{i^*}$ and $\vec{j^*}$ are the new basis as they are the result of the transformation of A on $\vec{i}$ and $\vec{j}$. The vector $\vec{b}$ is where we need $\vec{x}$ to land.

There are several ways to get the solution for that system, we will focus our attention on a technique called the Cramer's rule. As while it may not be the most efficient approach, it is the method with the best visual component. In this case, best means efficient in terms of computation. Our goal is not to learn how to calculate these solutions by hand. We have computers to do that. Instead, we just want to understand what is happening so we can better analyse the results.

Let's start by grabbing three elements from the plot above: the basis formed by the vectors $\vec{i}$ and $\vec{j}$, and the generic vector $\vec{x}$. Consider the following plot:

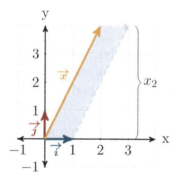

Figure 4.22: The first plot of Cramer's rule.

The vector $\vec{x}$ represents any vector that verifies a system of equations; its coordinates are (x_1, x_2). In this case, the blue parallelogram has an area equal to $1.x_2$, meaning that the value of this area is equal to x_2. Let's be rigorous here; it is not the area, but rather the signed area, because x_2 can be negative. We can make a similar case with the other coordinate in $\vec{x}$, a parallelogram with an area of x_1 instead of x_2:

But What About a Matrix?

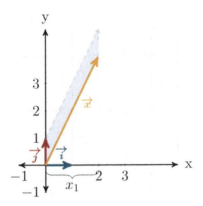

Figure 4.23: The second plot of Cramer's rule.

The signed area of the blue parallelogram will be $1.x_1$, resulting in x_1. These parallelograms can also be built with the transformed versions of $\vec{i}$, $\vec{j}$ and $\vec{x}$. Let's start with the vector $\vec{i^*}$ that has coordinates $(-5, -4)$. But instead of $\vec{x}$, we have $\vec{b}$, which is the result of transforming $\vec{x}$ with A:

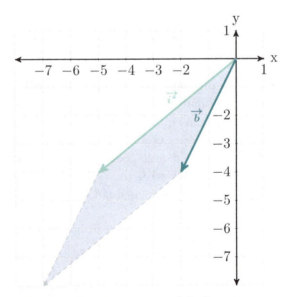

Figure 4.24: The solution of the linear system.

In the plots 4.22 and 4.23, we drew two different parallelograms. One with the vector $\vec{x}$ and $\vec{i}$, and the other formed by $\vec{x}$ and $\vec{j}$. These have areas values of x_2 and x_1 respectively. Following this, we transformed $\vec{i}$ and $\vec{x}$ to obtain a new parallelogram which we don't yet know the area of (as shown in the figure above). In order to derive x_2, one must find some relationship between the areas of the original and transformed parallelograms. Fortunately, we have a scalar that provides information relating to the magnitudes by which sizes are scaled via linear functions. So, if the area of the original parallelogram was x_2, this new shape will have an area of $det(A) \cdot x_2$:

$$Area = det(A) \cdot x_2$$

$$x_2 = \frac{Area}{det(A)}$$

We know the coordinates of vector $\vec{b}$ and we know where it lands, so we can create a new matrix:

$$\begin{pmatrix} -5 & -2 \\ -4 & -4 \end{pmatrix}$$

The determinant of this matrix will represent the area of the transformed parallelogram:

$$x_2 = \frac{\begin{vmatrix} -5 & -2 \\ -4 & -4 \end{vmatrix}}{\begin{vmatrix} -5 & 2 \\ -4 & 1 \end{vmatrix}} = \frac{(-5) \cdot (-4) - (-2) \cdot (-4)}{(-5) \cdot 1 - 2 \cdot (-4)} = \frac{12}{3} = 4$$

If we want to derive a formula to compute x_1 it is done in the same way as was shown for x_2. The only difference is that we will transform the vector $\vec{j}$ and use its transformed version alongside $\vec{b}$ to build the new parallelogram:

$$Area = det(A) \cdot x_1$$

$$x_1 = \frac{Area}{det(A)}$$

But What About a Matrix?

We know the coordinates of vector $\overrightarrow{b}$, so:

$$x_1 = \frac{\begin{vmatrix} -2 & 2 \\ -4 & 1 \end{vmatrix}}{\begin{vmatrix} -5 & 2 \\ -4 & 1 \end{vmatrix}} = \frac{(-2) \cdot 1 - (-2) \cdot (-4)}{3} = \frac{6}{3} = 2$$

We have found that the solution for our system is the vector $\overrightarrow{x} = (2,4)^T$. Indeed there are computers that can tally all of these tedious calculations, almost like having slaves without any feelings. I wonder what will happen if they ever get to the point of sentience. Will they miraculously channel all of their pain into the creation of new styles of dance and music that we can enjoy, even after the way we have treated them? Or will it be judgement day, and someone, John Connor style, will have to save us? I hope I haven't lost you because you're too young to have seen The Terminator? All the same, this technique is here to help you visualise what is happening. And, if you do choose to use a computer to solve a system, you won't think that these black boxes are performing magic. Also, this showcases a situation that often occurs in mathematics, in which there are several ways to solve the same problem. The task was to find a new vector that, when transformed, would land on a specific set of coordinates. The method for finding a solution was derived based on areas. Another way to solve a linear system is Gaussian elimination, which won't be covered here, though if you wish to learn it, I will say that you will have no problems with it.

There are still two things missing: we need the general formula for Cramer's rule, and we should verify what happens when we transform the vector $\overrightarrow{x}$ with the linear transformation A. First things first, Cramer's rule. Consider a linear system represented as follows:

$$A\overrightarrow{x} = \overrightarrow{b}$$

Where A is a squared matrix with size $n \times n$, $\overrightarrow{x} = (x_1, ..., x_n)^T$ and $\overrightarrow{b} = (b_1, b_2, ..., b_n)$. We can then define the values of the vector $\overrightarrow{x}$ by:

$$x_i = \frac{det(A_i)}{det(A)}$$

The value $det(A_i)$ is the matrix formed by replacing the i_{th} column of A with the column vector $\overrightarrow{b}$. Next, for the verification part, we have to transform $(2,4)^T$ with A and to check if we get $\overrightarrow{b}$:

$$\begin{pmatrix} -5 & 2 \\ -4 & 1 \end{pmatrix} \cdot \begin{pmatrix} 2 \\ 4 \end{pmatrix}$$

$$= \begin{pmatrix} -10+8 \\ -8+4 \end{pmatrix} = \begin{pmatrix} -2 \\ -4 \end{pmatrix}$$

Linear systems have many applications, but for now, I would like to explore four particular types of systems with some important characteristics. Consider a matrix $A_{m \times n}$ that maps vectors from $\mathbb{R}^n$ to vectors in $\mathbb{R}^m$. Let's also define two vectors, $\overrightarrow{x}$ and $\overrightarrow{y}$. The different systems that are worth paying special attention to are:

1. $A \cdot \overrightarrow{x} \neq 0$ row space of A.

2. $A \cdot \overrightarrow{x} = 0$ nullspace of A.

3. $A^T \cdot \overrightarrow{y} \neq 0$ column space of A.

4. $A^T \cdot \overrightarrow{y} = 0$ left nullspace of A.

The equations above define important subspaces in the field of linear algebra. A subspace is a space within another vector space where all the rules and definitions of vector spaces still apply. In case one, we are performing row-wise dot products with a vector $\overrightarrow{x}$, and our interest lies in the calculations performed between rows that are not perpendicular, so that the result is not zero. This subspace is called the row space. On the other hand, in equation number two, we want a subspace where every vector is transformed to zero. This is called the nullspace. Finally, cases three and four represent very similar concepts, but as we are now computing dot products with the transpose of A, these subspaces are for the columns instead of the rows. So, case three is the row space, while case four is the left nullspace.

There must be a relationship between the dimension of the space spanned by A and the subspaces that are part of A, namely those defined by the four systems shown above. To understand these relationships, we need to introduce another concept: rank. The rank of a matrix is the dimension of the vector space generated by its columns, meaning that it is also the basis of the column's space. Let's consider a matrix $A_{4 \times 3}$ defined as:

$$A = \begin{pmatrix} 1 & 2 & 0 \\ 1 & 2 & 1 \\ 2 & 4 & -1 \\ 3 & 6 & 0 \end{pmatrix}$$

To calculate the rank of A, we need to understand how many linearly independent columns exist in matrix A. This quantity will also give the number that defines the dimension of the column space. After a closer look, we can identify that the first column is half of the second, and there is no way of getting the third column by combining the first two, or by scaling any of them. So the matrix has a rank equal to two. There are two linearly independent vectors, $(1, 1, 2, 3)$ and $(0, 1, -1, 0)$, and they form a basis for the column space. Now, let's compute the nullspace of A. This space is such that $\{x \in \mathbb{R}^3 : A \cdot \overrightarrow{x} = \overrightarrow{0}\}$. We are looking for all of the values of $\overrightarrow{x}$ in $\mathbb{R}^3$, when transformed by A, result in the null vector, which is equivalent to:

$$\begin{pmatrix} 1 & 2 & 0 \\ 1 & 2 & 1 \\ 2 & 4 & -1 \\ 3 & 6 & 0 \end{pmatrix} \cdot \begin{pmatrix} x_1 \\ x_2 \\ x_3 \end{pmatrix} = \begin{pmatrix} 0 \\ 0 \\ 0 \end{pmatrix}$$

So it follows that:

$$\begin{cases} x_1 + 2x_2 = 0 \\ x_1 + 2x_2 + x_3 = 0 \\ 2x_1 + 4x_2 - x_3 = 0 \\ 3x_1 + 6x_2 = 0 \end{cases} \Leftrightarrow \begin{cases} x_1 = -2x_2 \\ x_3 = 0 \end{cases}$$

I am about to drop some more notation so brace yourselves.

The nullspace is represented by:

$$nullspace(A) = \{(c, -2c, 0) : c \in \mathbb{R}\}$$

And in this particular case, it can also be represented by:

$$Span\{(1, -2, 0)\}$$

The dimension of the nullspace, which is also called the $nullity(A)$, is equal to one, meaning that the basis of that space has only one vector. At this point, we have the dimensions of the column space and the nullspace: two and one respectively. Coincidentally or not, the dimension of the column space, which is equivalent to the rank of A plus the dimension of the nullspace, is equal to the number of columns in A. Obviously this is not a coincidence at all; the theorem of rank-nullity states that:

$$rank + nullity(A) = n \qquad (4.23)$$

The proof of this theorem is extensive, and I will omit it, but what we can take from it will be extremely useful. The rank of A plus its nullity equals n, the number of columns. Apologies, but we must introduce another simple definition at this time, for the term full rank. A matrix A of size $m \times n$ is said to be of full rank when the rank is equal to the $min(m, n)$. There are two possibilities to consider: $m \geq n$ and $m < n$. But why would we care about this? Well, it just so happens that we can understand the dimension of the nullspace when a matrix is of full rank by just looking at the order of magnitude of m and n. If A is of full rank and $m < n$, the rank of A is equal to m. By equation 4.23, we have $m + nullity(A) = n$, which is the same as:

$$nullity(A) = n - m \quad \text{when} \quad m < n \quad \text{and} \quad A \quad \text{is full rank.}$$

On the other hand, if $m \geq n$, it follows that the rank of A is equal to n, so the nullspace has a dimension of 0:

$$nullity(A) = n - n = 0 \quad \text{when} \quad m \geq n \quad \text{and} \quad A \quad \text{is full rank.}$$

If the matrix is square, then $n = m$, so it makes sense that a square matrix is of full rank when the rank of A is equal to n.

There are a couple more relationships between the nullity, the rank, and the determinant that are worth studying. Let A be such that:

$$A = \begin{pmatrix} \vec{v_1} & \vec{v_2} & ... & \vec{v_n} \end{pmatrix}$$

Where all the vectors $\vec{v}$ are the columns of A. With this, the equation for nullity can be expressed by:

$$A \cdot \vec{x} = \vec{0} \Leftrightarrow \begin{pmatrix} \vec{v_1} & \vec{v_2} & ... & \vec{v_n} \end{pmatrix} \cdot \begin{pmatrix} x_1 \\ x_2 \\ \vdots \\ x_n \end{pmatrix} = \begin{pmatrix} 0 \\ 0 \\ \vdots \\ 0 \end{pmatrix}$$

Which is the same as:

$$x_1 \cdot \vec{v_1} + x_2 \cdot \vec{v_2} + ... + x_n \cdot \vec{v_n} = \vec{0} \tag{4.24}$$

If we say that all the vectors are linearly independent, the only way for equation 4.24 to be verified is for all of the factors represented by the x's to be equal to 0. Otherwise, it will be possible to get at least one term as a function of another; consequently, the assumption of linear independence won't be valid. Therefore, if the nullspace is of dimension zero, it also means that the determinant is non-zero. As you can see, the rank provides a lot of information about linear transformations and is an excellent way of exploring some of their characteristics.

When introducing the rank and deducing its relationships and properties, we paid a lot of attention to the nullspace of a matrix. This space is also essential when it comes to introducing one more notion, the last one before we vault into the concepts that are often used in machine learning: the principal component analysis and the single value decomposition.

Even though we are already well-equipped in terms of linear algebra knowledge, the feeling in the air is that everything really starts here. Let's initiate this new part of the odyssey by defining a particular case of a linear system. What if we wish to find a vector $\vec{v}$ that, when transformed by A, lands on $\vec{v}$ or on a scaled version of it? Before, we were looking for a vector $\vec{x}$

that, when transformed, would land on a generic vector $\vec{b}$. We can define a scaling factor by λ and formulate the system with the following equation:

$$A\vec{v} = \lambda\vec{v} \tag{4.25}$$

4.3.1 Line Up Vectors! The Eigen Stuff

Good people of the sun, allow me to introduce you to the Eigen "stuff". These vectors $\vec{v}$ that, when transformed by a matrix, land on itself or on a scaled version of itself are called the eigenvectors. They happen to be like the police as they are always with someone else, so for each eigenvector, you will have a corresponding eigenvalue λ, the scalar by which the vector stretches or shrinks. To put it simply, think of a matrix as a machine that rotates and/or scales vectors. An eigenvector is a vector that, when put through this machine, only gets stretched or shrunk by a certain amount but does not get rotated:

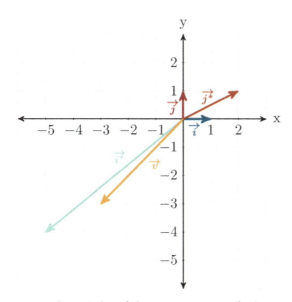

Figure 4.25: A reminder of the representation of a linear system.

In the plot above, $\overrightarrow{i^*}$, $\overrightarrow{j^*}$ represent the transformed version of the standard basis $\overrightarrow{i}$, $\overrightarrow{j}$. We are looking for one or more vectors that, when modified by A, land on a linear combination of $\overrightarrow{v}$:

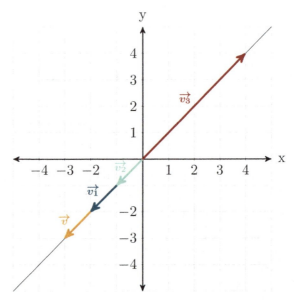

Figure 4.26: A mosh pit of vectors: some possibilities for eigenvectors.

We have a few examples of eigenvectors in the plot above, and, as they are all linear combinations of each other, they all land on the black line that represents that particular vector span. Let's manipulate equation 4.25 to get it into a format where the computation of both the eigenvectors and eigenvalues will be more straightforward. It is hard to find a good use case for manipulation. However, perpetrating this act to produce an equation still seems to be accepted. So far, nobody seems to care, and it comes with benefits for the mathematician. Let's exploit this:

$$A\overrightarrow{v} = \lambda\overrightarrow{v}$$

We've learned about the identity matrix on our journey into linear algebra, and so we can say:

$$A\overrightarrow{v} = \lambda I \overrightarrow{v}$$

Multiplying the identity matrix I by a given vector $\vec{v}$ will result in the vector $\vec{v}$. That equation is equivalent to:

$$A\vec{v} - \lambda I \vec{v} = \vec{0}$$

We can then write it as:

$$(A - \lambda I)\vec{v} = \vec{0} \qquad (4.26)$$

There are two ways for 4.26 to be verified. The first one is not very interesting (the trivial solution). This is when $\vec{v} = 0$. That told us nothing, so we need to find a $\vec{v}$ other then $\vec{v} = 0$. The element $(A - \lambda I)$ is a matrix whose diagonal elements are part of the diagonal of A minus the scalar λ, something like:

$$\begin{pmatrix} a & b \\ c & d \end{pmatrix} - \begin{pmatrix} \lambda & 0 \\ 0 & \lambda \end{pmatrix} = \begin{pmatrix} a - \lambda & b \\ c & d - \lambda \end{pmatrix}$$

So, equation 4.26 can be depicted as:

$$\underbrace{\begin{pmatrix} a - \lambda & b \\ c & d - \lambda \end{pmatrix}}_{R} \cdot \vec{v} = 0 \qquad (4.27)$$

Ideally, we would have an equation where we can apply some technique and calculate the $\lambda's$. If you recall, the determinant is a function that takes in a matrix and spits out a scalar, and we have a formula for it, so this might be an excellent path to follow to find an equation where we can isolate λ. However, the determinant alone won't be enough to get us to where we need to be. That zero on the right side of the equation causes constraints, and we need to consider them. Thankfully, there is a way to create a relationship between the determinant and zero. For example, consider the inverse of a matrix. For a matrix to have an inverse, its determinant has to be non-zero. In addition, it comes to mind that, if a matrix is invertible, the multiplication of this given matrix with its inverse results in the Identity matrix. Maybe we can work with this. Let's define the inverse of R by R^{-1} and replace this notation with 4.27:

$$R^{-1} \cdot R \cdot \vec{v} = R^{-1} \cdot \vec{0}$$

This results in:

$$I \cdot \vec{v} = \vec{0}$$

And that is as valuable as an ashtray on a motorbike. We specifically stated that $\vec{v} = 0$ was not of interest. But everything's not lost. What follows is that R must not be invertible, and like that, we guarantee any non-trivial solution! So, if R has no inverse, the determinant of that specimen must be zero:

$$\det(A - \lambda I) = 0$$

Geometrically, this also makes sense. We are transforming a vector $\vec{v}$ into a linear combination of itself. For that, the determinant has to be zero, as we saw previously. Let's follow an example and use A as the linear transformation:

$$\left| \begin{pmatrix} -5 & 2 \\ -4 & 1 \end{pmatrix} - \begin{pmatrix} \lambda & 0 \\ 0 & \lambda \end{pmatrix} \right| = 0$$

This is equivalent to:

$$\begin{vmatrix} -5 - \lambda & 2 \\ -4 & 1 - \lambda \end{vmatrix} = 0$$

$$(-5 - \lambda)(1 - \lambda) - 2 \cdot (-4) = 0$$

$$\lambda^2 + 4\lambda + 3 = 0$$

From our fundamental maths knowledge we know that the roots for that equation can be calculated with the following formula:

$$x = \frac{-b \pm \sqrt{b^2 - 4ac}}{2a}$$

Where:

$$ax^2 + bx + c = 0$$

Okay, so applying that formula to our equation gives us the following values for λ:

$$\lambda = -3 \quad \text{and} \quad \lambda = -1$$

We now have the scaling values that will stretch or squish the eigenvectors, but we still do not have the vectors. However, there is a formula that allows us to get these vectors: 4.26:

$$(A - \lambda I)\vec{v} = \vec{0}$$

Given that we know the λ values, we can compute the eigenvectors. Let's try with $\lambda = -3$:

$$\begin{pmatrix} -5 - (-3) & 2 \\ -4 & 1 - (-3) \end{pmatrix} \cdot \begin{pmatrix} v_1 \\ v_2 \end{pmatrix} = \vec{0}$$

$$\begin{pmatrix} -2 & 2 \\ -4 & 4 \end{pmatrix} \cdot \begin{pmatrix} v_1 \\ v_2 \end{pmatrix} = \vec{0}$$

Let's name that guy. Should we go for Francis? Maybe it is too much, but M should be enough:

$$M = \begin{pmatrix} -2 & 2 \\ -4 & 4 \end{pmatrix}$$

This represents a linear system, but the determinant of the matrix M is zero. Does this make sense? A determinant of 0? This means that we either have no solutions, or a lot of them. The latter is happening here. The coordinate values that verify that the system can be represented by vectors that land on the span of the vector that we are after. There is an implication to this determinant value; Cramer's rule will be doing a prophylactic isolation and can't show up. The system becomes:

$$\begin{cases} -2v_1 + 2v_2 = 0 \\ -4v_1 + 4v_2 = 0 \end{cases}$$

This is a simple system that is here as a showcase, therefore we can solve it with a "maneuver" like this:

$$\begin{cases} -2v_1 + 2v_2 = 0 \\ -4v_1 + 4v_2 = 0 \end{cases} \Leftrightarrow \begin{cases} -2v_1 = -2v_2 \\ -4v_1 + 4v_2 = 0 \end{cases}$$

$$\begin{cases} v_1 = v_2 \\ -4v_1 + 4v_2 = 0 \end{cases} \Leftrightarrow \begin{cases} v_1 = 1 \\ v_2 = 1 \end{cases}$$

So, the eigenvector corresponding to the eigenvalue $\lambda = -3$ is the vector $(1, 1)^T$, let's call it $\vec{t}$:

But What About a Matrix?

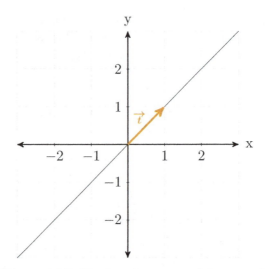

Figure 4.27: The first eigenvector of the example.

If we transform $\vec{t}$ with A, $\vec{t^*}$, the transformed version of $\vec{t}$ needs to land on that black line. It has to be the vector $(-3, -3)^T$; let's verify this:

$$\begin{pmatrix} -5 & 2 \\ -4 & 1 \end{pmatrix} \cdot \begin{pmatrix} -1 \\ -1 \end{pmatrix} = \begin{pmatrix} -3 \\ -3 \end{pmatrix}$$

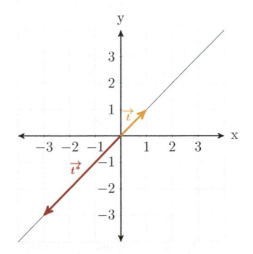

Figure 4.28: The transformation of a eigenvector.

Okay, that verifies the calculations. If we do something similar to that with $\lambda = -3$, but now with the other eigenvalue, $\lambda = -1$, we end up with:

$$\begin{pmatrix} -5 - (-1) & 2 \\ -4 & 1 - (-1) \end{pmatrix} \cdot \begin{pmatrix} v_1 \\ v_2 \end{pmatrix} = \overrightarrow{0}$$

$$\begin{pmatrix} -4 & 2 \\ -4 & 2 \end{pmatrix} \cdot \begin{pmatrix} v_1 \\ v_2 \end{pmatrix} = \overrightarrow{0}$$

I will omit some of the steps of the calculations as they are like the ones that we did for the other eigenvalue, but this time, the eigenvector is equal to $\overrightarrow{v} = (\frac{1}{2}, 1)^T$. Similarly, if we transform $\overrightarrow{v}$ into $\overrightarrow{v^*}$ making use of the linear transformation A, $\overrightarrow{v^*}$ has to land on the same line as $\overrightarrow{v}$. Let's verify:

$$\begin{pmatrix} -5 & 2 \\ -4 & 1 \end{pmatrix} \cdot \begin{pmatrix} \frac{1}{2} \\ 1 \end{pmatrix} = \begin{pmatrix} -\frac{1}{2} \\ -1 \end{pmatrix}$$

As expected, the resultant vector is a linear combination of the original vector that we transformed with the matrix A. Let's visualise this result:

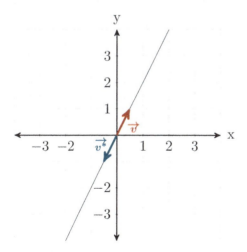

Figure 4.29: The second transformation of an eigenvector.

In essence, we found a particular group of vectors that, when transformed, land on its span. If you are doubtful about the utility of this concept, I empathise with that, but allow me to try to convince you that this was not an attempt to put a few more pages in this book so that it gets to at least 100 pages! So, let's bring back the 90's! Sorry, I mean matrix A, its eigenvalues, and the respective eigenvectors:

$$A = \begin{pmatrix} -5 & 2 \\ -4 & 1 \end{pmatrix}, \quad \lambda = -3, \quad \vec{t} = \begin{pmatrix} 1 \\ 1 \end{pmatrix}$$

$$\lambda = -1, \quad \text{and} \quad \vec{v} = \begin{pmatrix} \frac{1}{2} \\ 1 \end{pmatrix}$$

As a friendly reminder, this is how we can represent the system to calculate the eigenvalues:

$$A\vec{v} = \lambda\vec{v}$$

Okay, if we now replace $\vec{v}$ and λ with each eigenvalue and eigenvector pair, we can form two different equations:

$$A \begin{pmatrix} 1 \\ 1 \end{pmatrix} = -3 \begin{pmatrix} 1 \\ 1 \end{pmatrix}$$

$$A \begin{pmatrix} \frac{1}{2} \\ 1 \end{pmatrix} = -1 \begin{pmatrix} \frac{1}{2} \\ 1 \end{pmatrix}$$

(4.28)

So, we have two vector equations. By forming one matrix with the eigenvectors and another with the eigenvalues, we can combine the two equations in 4.28 into a single equation with matrices:

$$A \overbrace{\begin{pmatrix} 1 & \frac{1}{2} \\ 1 & 1 \end{pmatrix}}^{P} = \overbrace{\begin{pmatrix} 1 & \frac{1}{2} \\ 1 & 1 \end{pmatrix}}^{P} \overbrace{\begin{pmatrix} -3 & 0 \\ 0 & -1 \end{pmatrix}}^{\Lambda}$$

Algebraically we have:

$$AP = P\Lambda \qquad (4.29)$$

When we studied the inverse of a matrix, we concluded that for a matrix A^{-1} to be the inverse of A, $AA^{-1} = I$. So, if we want to isolate A on the left side of equation 4.29, we can multiply both sides of the equation by P^{-1}:

$$APP^{-1} = P\Lambda P^{-1}$$

Which becomes:

$$A = P\Lambda P^{-1} \qquad (4.30)$$

4.3.2 I've Got the Power, to Power - Matrix Diagonalization

So, we can define A as a product of three other matrices, including the eigen base P and its inverse P^{-1} which represent two rotations. On the other hand, the matrix Λ is a diagonal matrix with eigenvalues for elements. On this matrix are the scaling term of equation 4.30. So, what's happening here is that we are representing a matrix by two rotations and one scaling term. Since we have a linear transformation that is split into three different matrices, it is worth taking a look at the properties of these matrices.

One of them stands out: Λ, which is a diagonal matrix. The only time we dealt with a matrix of this type was when the identity I was involved in our calculations. It turns out that this class of matrices, the diagonal ones, has some pretty incredible properties that make operations much lighter and more manageable. For example, to calculate the determinant of a matrix with this form, one only needs to multiply the elements that form the diagonal. Let's consider A to be a diagonal matrix defined as:

$$A = \begin{pmatrix} a & 0 \\ 0 & b \end{pmatrix}$$

By definition, the determinant of A is:

$$\begin{vmatrix} a & 0 \\ 0 & b \end{vmatrix} = a \cdot b - 0 \cdot 0 = a \cdot b$$

A has two dimensions, and these are always simple cases. However, if we have matrices of higher dimensionality, the calculations for the determinant are now more straightforward than the ones for non-diagonal matrices. Let's now consider a matrix A with three dimensions:

$$A = \begin{pmatrix} a & 0 & 0 \\ 0 & b & 0 \\ 0 & 0 & c \end{pmatrix}$$

We know that we can pick a row or a column and apply the Laplace expansion that we previously defined to calculate the determinant. If you don't remember, there is no need to feel bad or even go look for it. Here it is again:

$$det(A) = \sum_{i=1}^{n} (-1)^{i+j} a_{ij} M_{ij}$$

As you can see, in this formula, an element of the matrix is multiplied by a minor, a_{ij}. As this is a diagonal matrix, the only non-zero elements are those on the diagonal when $j = i$, so the determinant will simply be:

$$det(A) = (-1)^{2i} a_{ii} M_{ii}$$

This is the same as multiplying all of the elements on the diagonal together. It is fair to assume that if we can simplify the calculation of the determinant for a diagonal matrix, then the computation of its inverse will also have a more straightforward formula, and it does! Previously we specified that:

$$A^{-1} = \frac{1}{det(a)} (adj(A))$$

If we have all the 0 entries excluding the diagonal, the matrix $adj(A)$, is equal to:

$$adj(A) = \begin{pmatrix} M_{11} & 0 & 0 \\ 0 & M_{22} & 0 \\ 0 & 0 & M_{33} \end{pmatrix}$$

This is equivalent to:

$$adj(A) = \begin{pmatrix} bc & 0 & 0 \\ 0 & ac & 0 \\ 0 & 0 & ab \end{pmatrix}$$

And if we divide that $adj(A)$ by the determinant:

$$A^{-1} = \frac{\begin{pmatrix} bc & 0 & 0 \\ 0 & ac & 0 \\ 0 & 0 & ab \end{pmatrix}}{abc}$$

This results in:

$$A^{-1} = \begin{pmatrix} \frac{1}{a} & 0 & 0 \\ 0 & \frac{1}{b} & 0 \\ 0 & 0 & \frac{1}{c} \end{pmatrix}$$

This simplifies the calculation of an inverse. So, we have to replace each non-zero element with its inverse. The last thing to mention about diagonal matrices relates to a scenario in which we want to compute multiplications. Let's define a generic 2×2 matrix X as:

$$X = \begin{pmatrix} x_{11} & x_{12} \\ x_{21} & x_{22} \end{pmatrix}$$

There are two different scenarios here: The first, matrix multiplication, is not commutative; however, if one of the matrices in the operation is diagonal, we can easily get the results. If X is multiplied by a diagonal on the left, it will change the matrix row-wise. Otherwise, if we are multiplying X by a diagonal matrix on the right, then this will have an impact column-wise. This will apply to any matrix size:

$$\begin{pmatrix} x_{11} & x_{12} \\ x_{21} & x_{22} \end{pmatrix} \cdot \begin{pmatrix} a & 0 \\ 0 & b \end{pmatrix} = \begin{pmatrix} ax_{11} & bx_{12} \\ ax_{21} & bx_{22} \end{pmatrix}$$

$$\begin{pmatrix} a & 0 \\ 0 & b \end{pmatrix} \cdot \begin{pmatrix} x_{11} & x_{12} \\ x_{21} & x_{22} \end{pmatrix} = \begin{pmatrix} ax_{11} & ax_{12} \\ bx_{21} & bx_{22} \end{pmatrix}$$

Secondly, and this will be the most potent fact about this diagonalization technique, we can simplify the multiplication between matrices. As the non-zero elements are all on the diagonal, powering these matrices comes with comfort, as it will only be necessary to power each component of the diagonal. Just to explain, when we say powering, we're referring to the process of multiplying a matrix by itself, which is the same as raising it by a power.

Okay, the reality is that two-by-two matrices are baby talk. They make clear-cut examples, but when do we have the opportunity to work with only two dimensions in the real world? If you wish to go on with machine learning, you will most likely encounter some big ass matrices, and on top of that, consider the fact that you have to power them with a large exponent. Diagonalization will then come in handy to save the day and the computation bill! We know that A can be represented by:

$$A = P\Lambda P^{-1}$$

Say that A is a $n \times n$ matrix where n is giant (no joke). Think of a matrix with information about the weather, like states of observable climate conditions, where the matrix entries are probabilities of moving from one state to another. You could model this with some type of sequential model where the states are the weather conditions on a given day. An example of this type of model is a Markov Chain (which we won't go into right now, but it will be covered on a different book). Under these conditions, there is a case for powering a matrix because it will allow you to understand the system at a given state. So, powering A to the t (where t is large) can be computed by multiplying A by itself t times, which is extremely expensive in terms of computation time . Imagine all the dot products that we'd have to do. But don't worry, we can make use of matrix diagonalization: instead of...

$$A^t = A \cdot A \cdot A...A \quad t \quad \text{times}$$

...we could try and do something different. Let's start by squaring A:

$$A^2 = P\Lambda P^{-1}P\Lambda P^{-1} \tag{4.31}$$

A was replaced by its diagonalization form, and we can simplify equation 4.31, as we know that $P.P^{-1}$ is equal to the identity I:

$$A^2 = P\Lambda\Lambda P^{-1}$$

The matrix Λ is diagonal, so it follows that $\Lambda.\Lambda$ is equal to Λ^2 and the equation becomes:

$$A^2 = P\Lambda^2 P^{-1}$$

Right, this seems simpler to compute. Let's check what we can do if we have A^3. Well, $A^3 = A^2 \cdot A$, which can be written thus:

$$A^3 = P\Lambda^2 P^{-1} \cdot P\Lambda P^{-1}$$

Following the same logic as before:

$$A^3 = P\Lambda^3 P^{-1}$$

Let's check for A^4... I am joking, but you see where this is going. If you want to power A to t, the formula is:

$$A^t = P\Lambda^t P^{-1}$$

Powering matrices is a common practice in machine learning. You probably won't have to write code for it. But still, I can assure you that when you are applying algorithms to data, it is likely that at some point, it will happen. Suppose we try to interpret the equation for diagonalization geometrically. In that case, we see that we can represent a linear transformation as one rotation, followed by a stretch, and then another rotation. Let's experiment: For that we'll need a volunteer, anyone? Anyone at all? Mr. $\vec{u}$, welcome back, sir.

Firstly we will transform it with A. Following this, we will take a longer route, going from P^{-1} to Λ, and finally to P. Okay, $\vec{u}$ has the coordinates $(1,3)^T$, and when mapped by A it results in:

$$A \cdot \vec{u} = \begin{pmatrix} -5 & 2 \\ -4 & 1 \end{pmatrix} \cdot \begin{pmatrix} 1 \\ 3 \end{pmatrix} = \begin{pmatrix} 1 \\ -1 \end{pmatrix}$$

The plot bellow illustrates the transformation A on $\vec{u}$, which we represent with $\vec{u^*}$:

But What About a Matrix?

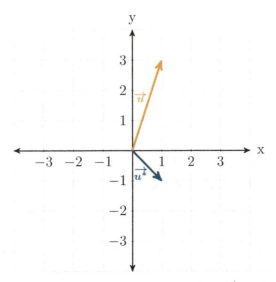

Figure 4.30: Matrix diagonalization - Where $\vec{v}$ needs to go.

Let's now check what happens if we rotate, stretch and rotate again the vector $\vec{u}$ with $P\Lambda P^{-1}$. The result has to be a vector with coordinates $(1, -1)^T$. There is one thing missing. Well, it is more like three; the dam matrices. We have calculated the eigenvectors of A, so we are good in regards to P. It is a matrix with the eigenvectors of A for columns. Now Λ is even more accessible. It is a matrix whose diagonal has for entries the eigenvalues of A, and the remaining entries are 0. That fellow P^{-1} is the inverse of P, and we already know how to invert a matrix, so we are in no danger of being stuck.

With this, we are ready to start, so bring on the first rotation, please, maestro:

$$P^{-1} \cdot \vec{u} = \begin{pmatrix} 2 & -1 \\ -2 & 2 \end{pmatrix} \cdot \begin{pmatrix} 1 \\ 3 \end{pmatrix} = \begin{pmatrix} -1 \\ 4 \end{pmatrix} = \vec{u_1^*}$$

Now, let's scale it:

$$\Lambda \cdot \vec{u_1^*} = \begin{pmatrix} -3 & 0 \\ 0 & -1 \end{pmatrix} \cdot \begin{pmatrix} -1 \\ 4 \end{pmatrix} = \begin{pmatrix} 3 \\ -4 \end{pmatrix} = \vec{u_2^*}$$

Finally, let's rotate it again:

$$P \cdot \vec{u_2^*} = \begin{pmatrix} 1 & \frac{1}{2} \\ 1 & 1 \end{pmatrix} \cdot \begin{pmatrix} 3 \\ -4 \end{pmatrix} = \begin{pmatrix} 1 \\ -1 \end{pmatrix} = \vec{u_3^*}$$

When we rotate $\vec{u}$ with P^{-1} we end up with a transformed vector $\vec{u_1^*}$ that is then stretched by λ into a new vector called $\vec{u_2^*}$. This is then rotated by P to result in $\vec{u_3^*}$ which in turn, has the same coordinates as $\vec{u}$. There is another name for this diagonalization technique. It is also known as eigendecomposition, which is a special case of matrix decomposition:

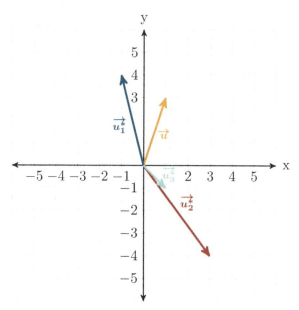

Figure 4.31: Matrix diagonalization - The full journey of $\vec{u}$.

The only catch is that the matrix that we decompose has to be square. Otherwise, we can't use this technique. Unfortunately, square matrices are as standard as tax rebates, which means they rarely come by, and, if you do come across one, you probably screwed up somewhere. Don't worry, it happens. So, if square matrices are not that common, there's got to be a way to decompose any matrix shape. And there is. You can refer to it as the single value decomposition, or by its street name, SVD.

But What About a Matrix?

These techniques of splitting a matrix into a multiplication of newly created matrices with particular characteristics have great visual interpretations. When it comes to introducing this new form for decomposing a matrix, we will focus on that.

Chapter 5

Break Them Down - Matrix Decomposition

We started simple, with a vector, and from that geometric idea we made our way through many linear algebra concepts. Some of them apply directly to real-world scenarios, like the dot product. Others are somewhat more abstract but still have great applicability to helping us deduce different concepts in linear algebra, like the eigenvectors. With the eigenvectors came a form of matrix decomposition, the eigendecomposition. We transformed a matrix into a product of three different matrices, namely two rotations and a scaling term. It is not a rule to decompose a matrix into three matrices. There are cases where we can decompose a matrix into a product of two others. However, in this book, we will cover two approaches to decomposition that happen to split a matrix into three different matrices: the eigendecomposition, which we just went through, and the single value decomposition, which will follow.

Before jumping into equations, let's check with some visuals what happens when we rotate, stretch and rotate a vector again. Consider a generic vector $\vec{t}$. Now to transform $\vec{t}$ into $\vec{t^*}$ we first perform a rotation of θ degrees from $\vec{t}$ to $\vec{t^*}$. The result is a vector called $\overrightarrow{t^{\theta}}$ that will then be scaled σ units into $\overrightarrow{t^{*}}$:

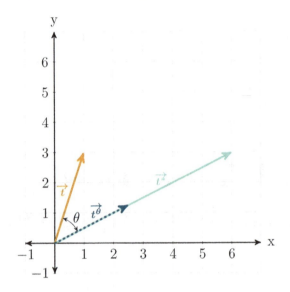

Figure 5.1: A rotation and a stretch.

In other words, we know where we need to go, we are just breaking the path down into different stages. If A is a linear transformation, we can rewrite A as a product of rotations and stretches. For example, if you wish to rotate a vector by θ degrees, a matrix like the following will do the job:

$$\begin{pmatrix} \cos(\theta) & -\sin(\theta) \\ \sin(\theta) & \cos(\theta) \end{pmatrix}$$

For a scaling operation, a diagonal matrix will be able to help:

$$\begin{pmatrix} \sigma_1 & 0 \\ 0 & \sigma_2 \end{pmatrix}$$

The first component of the vector gets scaled by a factor of σ_1, while the second gets treated by σ_2. So, a linear transformation will take any vector from the vector space, rotate it by the same angle, and then stretch it by some amount. Let's now focus our attention on a particular case, a circle.

We can define a pair of vectors to form a set of axes in the circle, and, as with the Cartesian plane, let's make them perpendicular:

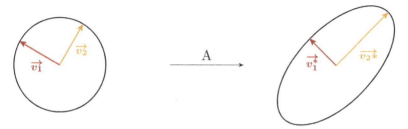

Figure 5.2: What happened to that circle?

The vectors $\vec{v_1}$ and $\vec{v_2}$ that form this axes are also of the length of the circle's radius. Maintaining the perpendicular orientation of these vectors after transforming them means that we will have a new axes in the newly transformed space. Therefore, it will probably be possible to derive the characteristics of these newly transformed vectors as a function of rotations and stretches.

If we stretch and rotate a circle, we end up with an ellipse, with $\vec{v_1}$ and $\vec{v_2}$ now constituting the minor and major axes of this newly mapped elliptical shape. In addition, if we define the norm of the vectors that compose the original axes to be one, perhaps we could understand how much they get scaled after a transformation, if we also introduce some parameters to reflect it. We have set up a good scenario in which to decompose a matrix into rotations and a stretch by doing these two things.

What we're missing is a linear transformation that not only guarantees that the angle between the vectors remains the same, but also preserves the magnitude of their norms. There is a type of matrix that always allows a transformation with these features to occur, namely the orthonormal matrix. But what are these orthonormal matrices? An orthonormal matrix, by definition, is a matrix whose columns are orthonormal vectors, meaning that the angle between them is 90 degrees and each of the vectors have a norm or length equal to one. These types of matrices have some excellent properties that make computations easier and very intuitive. Consider an example of a 2×2 orthonormal matrix. Let's call it Q: not Mr Q, just Q:

$$Q = \begin{pmatrix} 0 & 1 \\ -1 & 0 \end{pmatrix}$$

For the matrix to be orthonormal, each column vector has to be of length one and they need to be at 90 degrees to one another, which is the same as having a dot product equal to zero. Since we're still getting to know each other, and if you aren't confident that my police record is clean, then there is a case to say that our trust bond is still under development. So let's check the norms of those bad boys:

$$\vec{q_1} = (0, -1) \quad \text{and} \quad \vec{q_2} = (1, 0)$$

It follows that:

$$\|\vec{q_1}\| = \sqrt{0^2 + (-1)^2} = 1$$

$$\|\vec{q_2}\| = \sqrt{1^2 + 0^2} = 1$$

And the dot product is:

$$\vec{q_1} \cdot \vec{q_2} = \begin{pmatrix} 0 \\ -1 \end{pmatrix} \cdot \begin{pmatrix} 1 \\ 0 \end{pmatrix} = 0 \cdot 1 + (-1) \cdot 0 = 0$$

Okay, it seems we have an orthonormal basis; now I would like to do an experiment. The name orthonormal is the perfect setup for an adolescence defined by bullying. However, not everything is terrible. There is something else happening with these matrices. Multiplying Q^T by Q will create a relationship that we can leverage. We know that, for matrix multiplication, we use the dot product of the columns by the rows; if the vectors are orthogonal, some of those entries will be 0. Let's transpose Q and see what the deal is with this:

$$Q^T = \begin{pmatrix} 0 & -1 \\ 1 & 0 \end{pmatrix}$$

And:

$$Q^T Q = \begin{pmatrix} 0 & -1 \\ 1 & 0 \end{pmatrix} \cdot \begin{pmatrix} 0 & 1 \\ -1 & 0 \end{pmatrix} = \begin{pmatrix} 1 & 0 \\ 0 & 1 \end{pmatrix}$$

This is a result that is definitely worth paying attention to as we can build a connection between the inverse and the transpose matrices. But first, let's see if this happens for any size of matrix. For example, say that we have Q, which is an orthonormal matrix, but this time it has n vectors $\vec{q}$:

$$Q = \begin{pmatrix} | & | & & | \\ \vec{q_1} & \vec{q_2} & \cdots & \vec{q_n} \\ | & | & & | \end{pmatrix}$$

So $\vec{q_1}, ..., \vec{q_n}$ are all orthonormal vectors which form the columns of Q. The transpose of Q will have this shape:

$$Q^T = \begin{pmatrix} - & \vec{q_1}^T & - \\ - & \vec{q_2}^T & - \\ & \vdots & \\ - & \vec{q_n}^T & - \end{pmatrix}$$

It follows that, multiplying Q^T with Q:

$$Q = \begin{pmatrix} | & | & & | \\ \vec{q_1} & \vec{q_2} & \cdots & \vec{q_n} \\ | & | & & | \end{pmatrix} \cdot \begin{pmatrix} - & \vec{q_1}^T & - \\ - & \vec{q_2}^T & - \\ & \vdots & \\ - & \vec{q_n}^T & - \end{pmatrix} = \begin{pmatrix} 1 & 0 & \cdots & 0 \\ 0 & 1 & \cdots & 0 \\ \vdots & \vdots & \ddots & \vdots \\ 0 & 0 & \cdots & 1 \end{pmatrix}$$

The columns of Q are orthogonal vectors, so the dot products among them are 0. Therefore, if Q^T has the same vectors as Q, most of the dot products will also have zero value. The exception is when we take the dot product of the vector by itself, which means that all of the non-zero elements will be on the diagonal. Adding to this, we also know that the vectors $\vec{q}$ have a norm equal to one, so the resultant matrix will be the identity I. Let's bring back the transformation of a circle into an ellipse, but this time, let's define $\vec{v_1^*}$ and $\vec{v_2^*}$ in such a way that we can quantify how much they are scaled by A. The vectors $\vec{v_1}$ and $\vec{v_2}$, are scaled and rotated by A. We know that the angle between them will not change, as we're using an orthonormal matrix to perform the mapping, but how the vectors will scale at this point is still unknown. We can solve that by defining two unit vectors, $\vec{u_1}$ and $\vec{u_2}$, that will tell us the direction of the new set of axes formed by $\vec{v_1^*}$ and $\vec{v_2^*}$. Because $\vec{u_1}$ and $\vec{u_2}$ were defined with a norm that requires a length of 1, we can quantify the change in length that is a consequence of applying A. To do so, we can multiply them by a scalar, for example, σ_1 and σ_2. So the new axes will be:

$$\vec{v_1^*} = \sigma_1 \vec{u_1} \quad \text{and} \quad \vec{v_2^*} = \sigma_2 \vec{u_2}$$

We transformed a circle into an ellipse, and consequently, the vectors $\vec{v}$'s into $\vec{v^*}$'s. Then we decided to find a new way to represent the transformed version of these same $\vec{v}$'s. This allowed us to define these vectors by the multiplication of a scalar σ with a vector $\vec{u}$. There are names for all of these new appearing players. The $\vec{u}$'s are the principal axes and the σ's the single values.

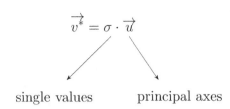

A transformation of $\vec{v_1}$ by the matrix A can then be represented as:

$$A \cdot \vec{v_1} = \sigma_1 \cdot \vec{u_1}$$

The only thing we did was to replace $\vec{v_1^*}$ with $A \cdot \vec{v_1}$. This equation reflects everything we have been discussing up until this point, a transformation of $\vec{v_1}$ enabled by A is equal to a scaled rotation of an orthonormal vector $\vec{u_1}$. The beauty of linear algebra is the ease with which we can move into higher dimensions.

Our example was a circle, which is in a vector space of size two. We selected this space so that we could visualise these concepts. In reality, if you apply any matrix decomposition, there is a significant chance that you will be working with larger space sizes. One of the applications of matrix decomposition is to reduce dimensions, so logically these matrices will have higher dimensionality. An explanation of this concept and a few others will follow shortly. We just need an equation, and we are halfway there. If you recall, the manner in which we handled earlier concepts was to define a starting case and then to accommodate a generalised solution for n. For all of the vectors, we can have an equation like:

$$A \cdot \vec{v_i} = \sigma_i \cdot u_i \quad \text{with} \quad i = 1, 2, ..., r \tag{5.1}$$

With:

$$\sigma_1 \geq \sigma_2 \geq ... \geq \sigma_r > 0$$

Algebraically, we can express it like:

$$\begin{pmatrix} \\ A \\ \\ \end{pmatrix} \cdot \begin{pmatrix} | & | & & | \\ \vec{v_1} & \vec{v_2} & ... & \vec{v_r} \\ | & | & & | \end{pmatrix} = \begin{pmatrix} | & | & & | \\ \vec{u_1} & \vec{u_2} & ... & \vec{u_r} \\ | & | & & | \end{pmatrix} \cdot \begin{pmatrix} \sigma_1 & ... & 0 \\ \vdots & \ddots & \vdots \\ 0 & ... & \sigma_r \end{pmatrix}$$

This is simply a different way of writing equation 5.1. Finally, we can use the following notation:

$$A \cdot V = U \cdot \Sigma$$

Greatness would come if A was by itself in this equation. In mathematics, loneliness is not necessarily a bad thing. We need to represent A via a product of three matrices, namely two rotations and a scaling term. But, we need A to be isolated, which we can achieve by multiplying each of the sides of the equation by the inverse of V:

$$A \cdot V \cdot V^{-1} = U \cdot \Sigma \cdot V^{-1}$$

And because V is orthonormal, the matrix $V^{-1} = V^T$:

$$A_{n \times m} = U_{m \times r} \cdot \Sigma_{r \times r} \cdot V_{r \times n}^T \qquad (5.2)$$

Where:

- U represents a rotation.

- Σ is the scaling matrix.

- V^T is another rotation.

The image below is a representation of the entire transformation and decomposition process:

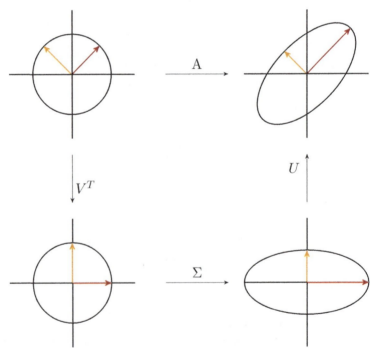

Figure 5.3: The detailed explanation of what went down with the circle - full transformation.

The vector's basis, or the axes, are rotated into a new set of orthonormal vectors via V^T. The resulting set of vectors is then scaled by the matrix Σ before being rotated by U to produce the desired version of the transformed vectors. Now would be a good moment to pause and look into the dimensionalities of the matrices in equation 5.2. The matrix U is so named because its columns are all the $\vec{u}$'s and these are all in $\mathbb{R}^m$. Meanwhile V^T has rows represented by the $\vec{v}$'s and these fellows are in $\mathbb{R}^n$. We can verify this by looking into, for example, the equation that we defined previously:

$$A_{m \times n} \cdot \vec{v_1} = \sigma_1 \cdot \vec{u_1}$$

If A is of size $m \times n$, the only way for that left-hand multiplication, $A_{m \times n} . \vec{v_1}$ to be possible is if $\vec{v_1}$ is a vector with dimensions $n \times 1$. The result of this will be a matrix of size $m \times 1$:

$$A_{m \times n} \cdot \vec{v}_{1_{n \times 1}} = \sigma_1 \cdot \vec{u_1}$$

Computing $A \cdot \vec{v_1}$ results in:

$$(A \cdot \vec{v_1})_{m \times 1} = \sigma_1 \cdot \vec{u_1}$$

If σ_1 is a scalar, it will have no effect on the dimensions of $\vec{u_1}$ after we multiply the two together. So, if the element on the left side of the equation has dimensions of $m \times 1$, then $\vec{u_1}$ must also have dimensions of $m \times 1$, otherwise we can't verify the equality above. Cool, so it follows that $v_1, ..., v_r \in \mathbb{R}^n$ and $u_1, ..., u_r \in \mathbb{R}^m$, meaning U is related to the columns in A and V is related to the rows. If we now consider the four subspaces that we introduced previously, we can define the column space and the row space of A as functions of U and V, such that:

- The set $u_1, .., u_r$ is a basis, in fact an orthonormal basis for the column space of A.

- The set $v_1, .., v_r$ is a orthonormal basis for the row space of A.

We have gathered some knowledge about all of the vectors $\vec{u}$'s and all of the vectors $\vec{v}$'s, but we still don't know what r

can possibly be. The matrix Σ is a diagonal matrix of size $r \times r$. Each column has only one non-zero value, and the position of this element varies from column to column. The fact that it is never at the same column index makes all the columns of Σ linearly independent. Consequently, Σ has a rank value of r. Going back to U, we know that it forms an orthonormal basis, and if this particular matrix has r columns, the rank of this matrix is also r. Multiplying U by Σ results in the matrix $U\Sigma$, which will also have rank r as Σ will only scale U. Therefore, the columns will remain linear and independent. For simplicity, let's refer to $U\Sigma$ as U^*. We now need to understand what happens when we multiply U^* by V^T; we still don't know how to relate the magnitude r to A. Okay, let's consider a generic column $\overrightarrow{y}$ that belongs to the column space of U^*V^T such that:

$$\overrightarrow{y} \in \text{Col}(U^*V^T)$$

So, by considering a column $\overrightarrow{y}$, we can take some vector $\overrightarrow{x} \in \mathbb{R}^n$ and transform this into $\overrightarrow{y}$ via U^*V^T:

$$\overrightarrow{y} = U^*V^T \cdot \overrightarrow{x} \qquad (5.3)$$

If we rewrite equation 5.3 to:

$$\overrightarrow{y} = U^* \left(V^T \cdot \overrightarrow{x} \right)$$

This indicates that $\overrightarrow{y}$ is also in the columns of U, which means that:

$$\overrightarrow{y} = U^* \cdot \overrightarrow{x^*} \qquad (5.4)$$

If $\overrightarrow{y}$ is in the column space of U, then we must also have a vector $\overrightarrow{x^*}$ such that $\overrightarrow{x^*} \in \mathbb{R}^r$, which when transformed by U^*, results in $\overrightarrow{y}$. If you recall, we have demonstrated that $V^T \cdot V$ is equal to the identity I. We can "squeeze" this into equation 5.4:

$$\overrightarrow{y} = U^* \cdot I_r \overrightarrow{x^*}$$

We can replace I_r by $V^T \cdot V$:

$$\overrightarrow{y} = U^*V^T \left(V \cdot \overrightarrow{x^*} \right)$$

Well, this means that this geezer (yes I lived in England and loved it!) $\vec{y} \in \text{Col}\left(U^* \cdot V^T\right)$. Before we got into this crazy sequence of equations, we had concluded that the rank of U was r, and now we have just shown that $\text{Col}\left(U^*\right) = \text{Col}\left(U^*V^T\right)$ so:

$$rank(U^*) = rank(U^*V^T)$$

If we wish to calculate the rank of A, we can do so using the following expression:

$$rank(A) = rank\left(U\Sigma V^T\right)$$

The conclusion is that the right side of the equation is of rank r, so we have:

$$rank(A) = r$$

Now we are in a state where we can understand the dimensions of the four subspaces of A. So far, we have defined two: the row space with size r, and the column space with the same dimension. The only subspaces that still haven't shown up to the party are the nullspace and the left nullspace. And believe me, we do need to know the sizes of these good fellows. For that, we can use the rank-nullity theorem. Let's start with the nullspace. We know that:

$$Rank(A) + Nullity(A) = n$$

So:

$$Nullity(A) = n - r$$

The nullspace of A has dimension $n - r$ and the basis for it are the vectors $v_{r+1}, ..., v_n$. The left nullspace is of size $m - r$ because:

$$Rank(A^T) + Nullity(A) = m$$

And its basis is $u_{r+1}, ..., u_m$. These are important results, and I will explain why, but first I would like to box them up so we can store them all in one place:

- $u_1, ..., u_r$ is an orthonormal basis for the column space and has dimensions of r.

- $u_{r+1}, ..., u_m$ is an orthonormal basis for the left nullspace and has dimensions of $m - r$.

- $v_1, ..., v_r$ is an orthonormal basis for the row space and has dimensions of r

- $v_{r+1}, ..., v_m$ is an orthonormal basis for the nullspace and has dimensions of $n - r$

Let's bring back the equation that reflects the decomposition of A:

$$A_{m \times n} = U_{m \times r} \Sigma_{r \times r} V_{r \times n}^T \tag{5.5}$$

If we take a closer look at the dimensions of each matrix, we can see that we are missing the vectors for both the nullspace and left nullspace in matrices V and U. Equation 5.5 represents the reduced version of the SVD. You may also remember this technique as the Single Value Decomposition that we spoke about earlier, but all the cool kids prefer to use its street name.

5.1 Me, Myself and I - The Single Value Decomposition (SVD)

The reduced form of this decomposition can be represented by:

$$A = U_r \Sigma_r V_r^T$$

Where r is the rank of matrix A as we proved earlier. A visual representation of the reduced single value decomposition is always useful:

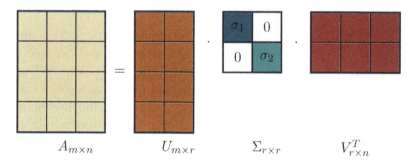

Figure 5.4: Colourful blocks that although not as cool as legos, are still helpful.

If we wish to have a full version of the singular value decomposition (SVD), we must incorporate the left nullspace in the matrix U_r and the nullspace in the matrix V_r. As these vectors are all orthogonal, we can do this without causing any problems with the orthonormality of U_r and V_r. In the matrix Σ we will include any single value that has a value of zero or, if necessary, add a row of zeros to ensure it's of size $n \times n$. We know the rank of U and the dimension of the nullspace, as these are r and $m - r$ respectively. These dimensions mean that U will be of size $m \times m$. Similarly, V will be of size $n \times n$ because the nullspace has dimensions of $n - r$. So, the equation of the full single value decomposition is very similar to that of the reduced version:

$$A_{m \times n} = U_{m \times m} \Sigma_{m \times n} V_{n \times n}^T \tag{5.6}$$

Graphically, this is what the full SVD looks like:

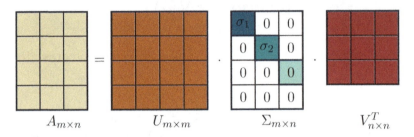

Figure 5.5: Even more colourful blocks! A failed attempt on coolness - Full SVD.

There is no information loss between the complete SVD and the reduced version of the same algorithm. The difference is that the reduced version is cheaper to compute. The SVD is probably the most potent decomposition from linear algebra that you can apply in machine learning. The algorithm has many applications, including recommendation systems, dimensionality reduction, creating latent variables, and computer vision. It is a versatile technique, and that is why it is crucial that we fully understand it.

The only thing we're missing, and it's probably one of the most essential, is an understanding of how to compute V, U, and Σ. We already did the visual work, so the central concept of this technique is covered in the plots. We defined the single value decomposition in such a way that the rotation matrices must be orthonormal. This is a constraint that will mean that the matrices from which we want to obtain a decomposition have to have some kind of properties that ensure orthogonality after some transformation. There is a type of matrix that will have something to do with orthogonality, the symmetrical ones. If a matrix is symmetrical, not only are the eigenvalues real but also the eigenvectors are orthogonal. Perhaps this is a good starting point, so let's explore it.

The definition of a symmetric matrix is straightforward. A symmetric matrix is such that $A = A^T$. Visually, it will be a matrix in which elements at the top of the diagonal will be equal to those at the bottom of the diagonal. Think of the diagonal as a mirror:

$$A = \begin{pmatrix} 2 & 3 & 6 \\ 3 & 4 & 5 \\ 6 & 5 & 9 \end{pmatrix}$$

If you switch the rows with the columns to get the transpose, you will end up with the same matrix. The only caveat now is that, in the SVD equation, A does not have to be a square matrix, let alone symmetrical. So, the conclusion is that the SVD is a load of crap and I am a liar... or... or it means that there is a way we a can transform A into a square symmetrical matrix. Let's start by thinking about the dimensions. If A is of size $m \times n$, then the only way to have a square matrix is if we multiply A by a matrix of size $n \times m$. Okay, so which matrix can we define to achieve such a thing? Let's bring back the SVD equation:

$$A = U\Sigma V^T \tag{5.7}$$

At this point, we only know A from equation 5.7. So randomly selecting a matrix of the correct size to multiply by A seems like a bad idea (right?!). After all the work we've invested into defining orthonormal transformations and scaling factors to keep track of movement in space, throwing in something random at this stage won't work. Like any subject politicians try to convince us about, we have two options to choose, meaning that if we select one, the other is automatically wrong.

No need to think about anything! It is either A or A^T. Picking A won't work because this matrix can be rectangular and therefore we can't multiply it by itself. So A^T is what remains. Remember that the multiplication of two matrices is a composition of scaled rotations achieved by dot products between rows and columns. A and A^T have the same vectors for entries, and the only difference is that they're displayed differently. So, we can guarantee that the matrix will be square, but we are still unsure if it will be symmetrical. Consider an example of matrix A defined as such, and let's check what happens when we perform the dot product between the rows and columns with the same colour:

$$A = \begin{pmatrix} 2 & 1 & 7 & 4 \\ 3 & 5 & 2 & 1 \\ 5 & -1 & 4 & 3 \end{pmatrix}_{3\times4} \quad \text{and} \quad A^T = \begin{pmatrix} 2 & 3 & 5 \\ 1 & 5 & -1 \\ 7 & 2 & 4 \\ 4 & 1 & 3 \end{pmatrix}_{4\times3}$$

Let's say B is the resultant matrix from the multiplication of A by A^T. B has dimensions of 3×3. The element $b_{1,2}$ is the dot product between the vectors highlighted in blue, and the element $b_{2,1}$ is the result of the dot product between the orange highlighted vectors. The result of these dot products is equal. The vectors are the same in the two operations, therefore $b_{1,2} = b_{2,1}$. This result will occur for every element of these matrices except for the diagonal since the diagonal comes after the dot product between two equal vectors. The first step of getting the three components of the SVD for the equation, $A = U \cdot \Sigma \cdot V^T$, is done: now we will multiply both sides by A^T:

$$A^T \cdot A = (U \cdot \Sigma \cdot V^T)^T \cdot U \cdot \Sigma \cdot V^T$$

There are a lot of T's in that equation, but as we have diagonal and orthonormal matrices, we can make the equation look a little better. It follows that $(A \cdot B)^T = B^T \cdot A^T$, and if we apply this to the equation, it becomes:

$$A^T \cdot A = V \cdot \Sigma^T \cdot U^T \cdot U \cdot \Sigma \cdot V^T$$

Okay, so I have two pieces of information about these matrices that will allow some further simplification:

1. That guy, Σ is a scaling matrix and therefore a diagonal matrix. The only non-zero entries are on the diagonal, and therefore the transpose is equal to the original matrix, i.e, $\Sigma = \Sigma^T$.

2. The matrix U is orthonormal, and it was proved earlier that the transpose of matrices with this property are equal to the inverse: $U^{-1} = U^T$.

It follows that:

$$A^T \cdot A = V \cdot \Sigma^T \cdot U^T \cdot U \cdot \Sigma.V^T$$

And the multiplication of U^T by U will result in the identity matrix such that:

$$A^T \cdot A = V \cdot \Sigma \cdot \Sigma \cdot V^T$$

Which is equivalent to:

$$A^T \cdot A = V \cdot \Sigma^2 \cdot V^T \tag{5.8}$$

We are getting closer to a way to compute V, which is not bad, but we still don't know how to get Σ or U. The problem is that we are depending on Σ to compute V. Our solution is to fetch something that we learned earlier and then to apply it to see if we can make some progress. Looking at 5.8, we know that Σ^2 is full of scalars on its diagonal, and, just to satisfy your curiosity, they are all positive. The only equation I can recall that we've covered lately with scalars on the diagonal of a matrix is the one we used to compute eigenvalues. Let's see if something can be done with this, but allow me to refresh your memory:

$$A \cdot \vec{v} = \lambda \cdot \vec{v} \tag{5.9}$$

Let's try to make equation 5.8 look similar to equation 5.9. For that, we need a vector or another matrix on the right side. We know that $V^T = V^{-1}$ and we can leverage this by multiplying 5.8 by V on both sides:

$$A^T \cdot A \cdot V = V \cdot \Sigma^2 \cdot V^{-1} \cdot V$$

And, it follows that:

$$A^T \cdot A \cdot V = V \cdot \Sigma^2 \tag{5.10}$$

Take a good look at this bad boy. Sometimes mathematics is almost like an illusion. Let's say that A is equal to $A^T \cdot A$, then V takes the place of $\vec{v}$, and the matrix which just has elements on its diagonal replaces λ. This means we have created an eigenvalue problem where Σ is an eigenvalue matrix. Okay, we can work with this! So, we have found a way to compute Σ and V, but we still need U. I multiplied A on the left by A^T, and all the U's went for a walk: they disappeared from the equation. So perhaps I can multiply A by A^T, but this time on the right. And who will be taking a walk this time? That's right; it will be the V's. Let's check:

$$A \cdot A^T = U \cdot \Sigma \cdot V^T (U \cdot \Sigma \cdot V^T)^T$$

I think you can see where this is going. We do the same as we did before when we computed the equation by multiplying A by A^T on the right:

$$A \cdot A^T = U \cdot \Sigma \cdot V^{-1} \cdot V\Sigma \cdot U^T$$

Then:

$$A \cdot A^T = U \cdot \Sigma^2 \cdot U^T$$

And multiplying by U on both sides, it follows that:

$$A \cdot A^T U = U \cdot \Sigma^2 \qquad (5.11)$$

We end up with a similar eigenvalue problem, and Σ happens to be the same for the two equations. Solving these eigenvalue equations will give us U and V. And as $A \cdot A^T$ is a square symmetric matrix, its eigenvalues will all be distinct and real; when root squared, they will give us the single values. The eigenvectors that form U and V will be orthogonal, allowing any vectors transformed by these matrices to preserve the angles between them. If we recall the "requirements" for all those transformations when done graphically, this was exactly it. There is only one more thing, the matrix Σ which contains the single values will be sorted such that, $\sigma_1 \geq \sigma_2 \geq ... \geq \sigma_n$.

I believe that the computation of such matrices will become solidified for you with a simple numerical example. Consider the matrix A defined as:

$$A = \begin{pmatrix} 1 & -2 & 0 \\ 0 & -2 & 1 \end{pmatrix} \quad \text{and} \quad A^T = \begin{pmatrix} 1 & 0 \\ -2 & -2 \\ 0 & 1 \end{pmatrix} \qquad (5.12)$$

The first step will be to compute $A \cdot A^T$ and $A^T \cdot A$:

$$A \cdot A^T = \begin{pmatrix} 1 & -2 & 0 \\ 0 & -2 & 1 \end{pmatrix} \cdot \begin{pmatrix} 1 & 0 \\ -2 & -2 \\ 0 & 1 \end{pmatrix} = \begin{pmatrix} 5 & 4 \\ 4 & 5 \end{pmatrix}$$

$$A^T \cdot A = \begin{pmatrix} 1 & 0 \\ -2 & -2 \\ 0 & 1 \end{pmatrix} \cdot \begin{pmatrix} 1 & -2 & 0 \\ 0 & -2 & 1 \end{pmatrix} = \begin{pmatrix} 1 & -2 & 0 \\ -2 & 8 & -2 \\ 0 & -2 & 1 \end{pmatrix}$$

To compute the matrix U, we need to solve the eigen problem in equation 5.11, let's start:

$$\det(AA^T - \lambda \cdot I) = \begin{vmatrix} 5 - \lambda & 4 \\ 4 & 5 - \lambda \end{vmatrix} = \lambda^2 - 10\lambda + 9$$

$$= (\lambda - 9)(\lambda - 1)$$

For roots we have $\lambda_1 = 9$ and $\lambda_2 = 1$. Now we need the eigenvectors:

$$A \cdot A^T \cdot \vec{u} = \lambda \cdot \vec{u}$$

For the value of $\lambda_1 = 9$ it follows that:

$$\begin{pmatrix} -4 & 4 \\ 4 & -4 \end{pmatrix} \cdot \begin{pmatrix} u_{11} \\ u_{12} \end{pmatrix} = 0$$

That, in turn is equivalent to:

$$\begin{cases} -4u_{11} = -4u_{12} \\ 4u_{11} = 4u_{12} \end{cases} \Leftrightarrow \begin{cases} u_{11} = u_{12} \end{cases}$$

So we have $u_{11} = 1$ and $u_{12} = 1$ meaning that $\vec{u_1} = (1,1)^T$. Remember that U is an orthonormal basis. Its vectors must have length 1. We need to normalise them. Let's start by calculating $\vec{u_1}$'s norm:

$$\|u_1\| = \sqrt{1 + 1} = \sqrt{2}$$

So u_1 is such that:

$$\vec{u_1} = \begin{pmatrix} \frac{1}{\sqrt{2}} \\ \frac{1}{\sqrt{2}} \end{pmatrix}$$

Doing the same for the eigen value, $\lambda = 1$:

$$\begin{pmatrix} 4 & 4 \\ 4 & 4 \end{pmatrix} \cdot \begin{pmatrix} u_{21} \\ u_{22} \end{pmatrix} = 0$$

With:

$$\left\{ 4u_{21} = -4u_{22} \right. \Leftrightarrow \left\{ u_{21} = -u_{22} \right.$$

In this case we have found that $u_{21} = -1$ and $u_{22} = 1$, meaning that $\vec{u_2} = (-1, 1)^T$. The norm of $\vec{u_2}$ is also $\sqrt{2}$, therefore the normalised form of $\vec{u_2}$ is:

$$\vec{u_2} = \begin{pmatrix} -\frac{1}{\sqrt{2}} \\ \frac{1}{\sqrt{2}} \end{pmatrix}$$

Perfect, so we have computed the matrix U. It is an orthonormal matrix because the eigenvectors have been normalized and $A^T \cdot A$ is a symmetrical matrix. Its eigenvectors are therefore orthogonal:

$$U = \begin{pmatrix} \frac{1}{\sqrt{2}} & -\frac{1}{\sqrt{2}} \\ \frac{1}{\sqrt{2}} & \frac{1}{\sqrt{2}} \end{pmatrix}$$

Next, we need the eigenvalues of $A.A^T$:

$$\det(A^T \cdot A - \lambda.I) = \begin{vmatrix} 1 - \lambda & -2 & 0 \\ -2 & 8 - \lambda & -2 \\ 0 & -2 & 1 - \lambda \end{vmatrix}$$

$$= (1 - \lambda)\left[(8 - \lambda)(1 - \lambda) - 4\right] - 4(1 - \lambda))$$

$$= (1 - \lambda)^2(8 - \lambda) - 8(1 - \lambda)$$

$$= (1 - \lambda)(8 - 9\lambda + \lambda^2 - 8)$$

$$= \lambda(1 - \lambda)(\lambda - 9)$$

Thus, it follows that, $\lambda_1 = 9$, $\lambda_2 = 1$ and $\lambda_3 = 0$. And to define V, we are just missing the normalised eigenvector for each eigenvalue:

For $\lambda_1 = 9$:

$$\begin{pmatrix} -8 & -2 & 0 \\ -2 & -1 & -2 \\ 0 & -2 & -8 \end{pmatrix} \cdot \begin{pmatrix} v_{11} \\ v_{12} \\ v_{13} \end{pmatrix} = 0$$

$$\begin{cases} -8v_{11} - 2v_{12} = 0 \\ -2v_{11} - v_{12} - 2.v_{13} = 0 \\ -2v_{12} - 8v_{13} = 0 \end{cases} \Leftrightarrow \begin{cases} -8v_{11} = 2v_{12} \\ -2v_{11} - v_{12} - 2v_{13} = 0 \\ -2v_{12} = 8v_{13} \end{cases}$$

$$\begin{cases} -8v_{11} = -8v_{13} \\ -2v_{11} - v_{12} - 2v_{13} = 0 \end{cases} \Leftrightarrow \begin{cases} v_{11} = v_{13} \\ -2v_{11} - v_{12} - 2v_{11} = 0 \end{cases}$$

$$\begin{cases} v_{11} = v_{13} \\ -v_{12} - 4v_{11} = 0 \end{cases} \Leftrightarrow \begin{cases} v_{11} = v_{13} \\ v_{12} = 4v_{11} \end{cases}$$

For $\overrightarrow{v_1}$ we have the vector $(1, -4, 1)^T$, and its norm is given by:

$$\|v_1\| = \sqrt{1^2 + (-4)^2 + 1^2} = \sqrt{18}$$

This makes the normalised version of $\overrightarrow{v_1}$ equal to:

$$\overrightarrow{v_1} = \begin{pmatrix} \frac{1}{\sqrt{18}} \\ \frac{-4}{\sqrt{18}} \\ \frac{1}{\sqrt{18}} \end{pmatrix}$$

Carrying on with the eigenvalue calculations, for $\lambda_2 = 1$, we have the following:

$$\begin{pmatrix} 0 & -2 & 0 \\ -2 & 7 & -2 \\ 0 & -2 & 0 \end{pmatrix} \cdot \begin{pmatrix} v_{21} \\ v_{22} \\ v_{23} \end{pmatrix} = 0$$

$$\begin{cases} -2v_{22} = 0 \\ -2v_{21} + 7v_{22} - 2v_{23} = 0 \\ -2v_{22} = 0 \end{cases} \Leftrightarrow \begin{cases} v_{22} = 0 \\ -2v_{21} - 2v_{23} = 0 \end{cases}$$

$$\begin{cases} v_{22} = 0 \\ v_{21} = -v_{23} \end{cases}.$$

For $\vec{v_2}$ we have the vector $\vec{v_2} = (-1, 0, 1)^T$ and the norm of $\vec{v_2}$:

$$\|v_2\| = \sqrt{(-1)^2 + 1^2} = \sqrt{2}$$

So the normalised version of $\vec{v_2}$ is:

$$\vec{v_2} = \begin{pmatrix} -\frac{1}{\sqrt{2}} \\ 0 \\ \frac{1}{\sqrt{2}} \end{pmatrix}$$

Finally, $\vec{v_2}$, where $\lambda_3 = 0$, is:

$$\begin{pmatrix} 1 & -2 & 0 \\ -2 & 8 & -2 \\ 0 & -2 & 1 \end{pmatrix} \cdot \begin{pmatrix} v_{31} \\ v_{32} \\ v_{33} \end{pmatrix} = 0$$

$$\begin{cases} v_{31} - 2v_{32} = 0 \\ -2v_{31} + 8v_{32} - 2v_{33} = 0 \\ -2v_{32} + v_{33} = 0 \end{cases} \Leftrightarrow \begin{cases} v_{31} = 2v_{32} \\ 2v_{32} = 2v_{33} \end{cases}$$

$$\begin{cases} v_{31} = 2v_{32} \\ v_{31} = v_{33} \end{cases}$$

For $\vec{v_3}$ we have the vector $(2, 1, 2)^T$, and its norm is equal to:

$$\sqrt{2^2 + 1^2 + 2^2} = \sqrt{9} = 3$$

Which makes the normalised version of $\vec{v_3}$:

$$\vec{v_3} = \begin{pmatrix} \frac{2}{3} \\ \frac{1}{3} \\ \frac{2}{3} \end{pmatrix}$$

The only matrix we're missing is Σ, but in fact, we have all the information we need to populate this matrix because the σ's are the square root of the eigenvalues, so $\sigma_i = \sqrt{\lambda_i}$:

$$\Sigma = \begin{pmatrix} 3 & 0 & 0 \\ 0 & 1 & 0 \end{pmatrix}$$

That was quite a bit of work, but thankfully we have computers, so it's not necessary to do this every time you wish to apply this algorithm. All the same, I believe that going over a simple example is always helpful to solidify your understanding of the algorithm in full. There's also the added bonus that these three matrices will allow you to do several things, so understanding how they are derived and what constitutes each is a great piece of knowledge to add to your arsenal. Finally, we can represent A as:

$$U\Sigma V^T = \begin{pmatrix} \frac{1}{\sqrt{2}} & -\frac{1}{\sqrt{2}} \\ \frac{1}{\sqrt{2}} & \frac{1}{\sqrt{2}} \end{pmatrix}_{2\times2} \cdot \begin{pmatrix} 3 & 0 & 0 \\ 0 & 1 & 0 \end{pmatrix}_{2\times3} \cdot \begin{pmatrix} \frac{1}{\sqrt{18}} & -\frac{4}{\sqrt{18}} & \frac{1}{\sqrt{18}} \\ -\frac{1}{\sqrt{2}} & 0 & \frac{1}{\sqrt{2}} \\ \frac{2}{3} & \frac{1}{3} & \frac{2}{3} \end{pmatrix}_{3\times3}$$

We should check that I'm not just talking nonsense or making stuff up! Every time I use a sauna, I think about something similar. What if the guy that invented this concept of a room that gets really warm and steamy was just messing with us, and now everybody thinks it's great? To avoid this kind of train of thought, let's multiply those three matrices:

$$= \begin{pmatrix} \frac{3}{\sqrt{2}} & -\frac{1}{2} & 0 \\ \frac{3}{\sqrt{2}} & \frac{1}{2} & 0 \end{pmatrix} \cdot \begin{pmatrix} \frac{1}{\sqrt{18}} & -\frac{4}{\sqrt{18}} & \frac{1}{\sqrt{18}} \\ -\frac{1}{\sqrt{2}} & 0 & \frac{1}{\sqrt{2}} \\ \frac{2}{3} & \frac{1}{3} & \frac{2}{3} \end{pmatrix}$$

$$= \begin{pmatrix} 1 & -2 & 0 \\ 0 & -2 & 1 \end{pmatrix} = A$$

In the example we've just covered, we computed the full version of the single value decomposition, which implies that both the null and left nullspaces are part of the matrices V^T and U. Previously, we concluded that the nullspace would be in the rows of matrix V^T. The equation that allowed us to compute the eigenvector for an eigenvalue of 0 is the same as the one we defined for calculating the nullspace: $A \cdot \vec{v} = 0$. But what about the left nullspace? Well, as a result of the decomposition, I can tell you that this is $(0,0)$, but let's verify this. We know that the left null is:

$$A^T \cdot \vec{x} = 0 \Leftrightarrow \begin{pmatrix} 1 & 0 \\ -2 & -2 \\ 0 & 1 \end{pmatrix} \cdot \begin{pmatrix} x_1 \\ x_2 \end{pmatrix} = \begin{pmatrix} 0 \\ 0 \end{pmatrix}$$

If we solve that system, we end up with:

$$\begin{cases} x_1 = 0 \\ -2x_1 - 2x_2 = 0 \\ x_2 = 0 \end{cases}$$

As expected, the left nullspace is the vector $(0,0)$, and its dimension is zero. Therefore, if we wish to use the reduced decomposition version, we need to remove the nullspace from the matrix V and adjust the matrix Σ accordingly:

$$U_r \Sigma_r V_r^T = \begin{pmatrix} \frac{1}{\sqrt{2}} & -\frac{1}{\sqrt{2}} \\ \frac{1}{\sqrt{2}} & \frac{1}{\sqrt{2}} \end{pmatrix}_{2\times2} \cdot \begin{pmatrix} 3 & 0 \\ 0 & 1 \end{pmatrix}_{2\times2} \cdot \begin{pmatrix} \frac{1}{\sqrt{18}} & -\frac{4}{\sqrt{18}} & \frac{1}{\sqrt{18}} \\ -\frac{1}{\sqrt{2}} & 0 & \frac{1}{\sqrt{2}} \end{pmatrix}_{2\times3}$$

This is a lot of mathematical contortions: rotations, stretches, and equations. Next, I'll present a small example of a real-life scenario where the decomposition of matrices is an excellent way to extract practical, valuable information. But before that, I would like to point out that the single value decomposition includes nearly everything we've learned so far in this book: it is thus an excellent way for us to make a self diagnosis on the solidity of our understanding of the linear algebra concepts that have been presented.

On the theme of applicability, I believe that there are three main areas in which to use SVD. The first one, my favourite, is the creation of latent variables. These are hidden relationships that exist within data. An elementary example would be for me to predict if it was raining or not without information on the weather report. For instance, I could do so by observing whether your jacket is wet or not when you arrive home.

I can also discuss an example where I produced great results when trying to predict outcomes related to human emotions, all by analysing hidden variables. It's hard to have a data set containing a happiness feature from the get-go, but perhaps these can be extracted from the data as a latent variable. With SVD, these variables are present in the matrices U and V. These types of variables also allow us to reduce the dimensionality of a data set.

Secondly, the singular value decomposition allows us to make an excellent approximation of a matrix. An approximation can be helpful for two purposes: we can remove data or information that is not relevant to what we are trying to do, making a prediction, for example; and we can condense this approximation to a few vectors and scalars, which can be extremely useful for storing data more efficiently.

Lastly, you can build a simple recommendation system. As this is linear algebra and the focus of this book is to visualise concepts in order to understand how they work, I will illustrate all of the applications described above in a real-world example. Let's consider a matrix containing item ratings by users. Each user can rate a given item with an integer score that ranges from 1 to 5. A matrix is an excellent choice to represent these dynamics. Let's say we have six items and we'll say they are songs. Yep, six different songs that users have rated according to their preferences, and our sample has eight users. Now yes, I know, this is a tiny, tiny data sample. And yes, you are likely to encounter far bigger data sets when the time comes to apply a single value decomposition, but for illustration purposes, this is enough. Let's then construct this matrix:

$$A = \begin{array}{c} \\ user_0 \\ user_1 \\ user_2 \\ user_3 \\ user_4 \\ user_5 \\ user_6 \\ user_7 \end{array} \begin{array}{cccccc} song_0 & song_1 & song_2 & song_3 & song_4 & song_5 \\ \left(\begin{array}{cccccc} 1 & 2 & 3 & 4 & 1 & 5 \\ 4 & 1 & 0 & 3 & 5 & 1 \\ 3 & 0 & 1 & 2 & 2 & 4 \\ 0 & 2 & 1 & 0 & 1 & 5 \\ 0 & 3 & 1 & 3 & 4 & 1 \\ 2 & 1 & 1 & 4 & 2 & 3 \\ 0 & 4 & 1 & 4 & 0 & 3 \\ 5 & 0 & 3 & 0 & 1 & 5 \end{array} \right) \end{array} \quad (5.13)$$

The matrix A is defined by 5.13. Now we need to decompose this into $U\Sigma V^T$, but this time I will use a computer to calculate the full version of the SVD. A will then be decomposed into:

$$A_{8\times6} = U_{8\times8} \cdot \Sigma_{8\times6} \cdot V_{6\times6}^T$$

It follows that the matrix U has the following entries:

	$context_1$	$context_2$	$context_3$	$context_4$	$context_5$	$context_6$	$context_7$	$context_8$
$user_0$	−0.45	0.03	−0.34	−0.26	−0.13	−0.55	0.02	−0.52
$user_1$	−0.35	0.17	0.70	0.08	0.06	0.26	−0.13	−0.49
$user_2$	−0.35	−0.23	0.14	−0.06	−0.46	0.17	0.71	0.19
$user_3$	−0.27	−0.15	−0.37	0.72	−0.28	0.25	−0.25	−0.10
$user_4$	−0.28	0.51	0.13	0.44	0.21	−0.45	0.18	0.39
$user_5$	−0.36	0.12	0.07	−0.36	−0.37	0.04	−0.57	0.48
$user_6$	−0.32	0.37	−0.42	−0.23	0.42	0.55	0.16	0.02
$user_7$	−0.38	−0.68	0.09	0.00	0.55	−0.10	−0.09	0.19

$U =$ (to the left of the matrix above)

For a full version of V, we have:

$$V = \begin{array}{c} \\ song_0 \\ song_1 \\ song_2 \\ song_3 \\ song_4 \\ song_5 \end{array} \begin{array}{cccccc} context_1 & context_2 & context_3 & context_4 & context_5 & context_6 \\ \left(\begin{array}{cccccc} -0.36 & -0.28 & -0.26 & -0.46 & -0.35 & -0.62 \\ -0.46 & 0.45 & -0.20 & 0.54 & 0.28 & -0.41 \\ 0.55 & -0.31 & -0.19 & 0.00 & 0.61 & -0.44 \\ -0.28 & 0.33 & -0.10 & -0.66 & 0.56 & 0.23 \\ 0.40 & 0.64 & 0.42 & -0.24 & -0.20 & -0.40 \\ 0.34 & 0.34 & -0.82 & -0.02 & -0.27 & 0.17 \end{array} \right) \end{array}$$

Finally, Σ is equal to:

$$\Sigma = \begin{pmatrix} 15.72 & 0 & 0 & 0 & 0 & 0 \\ 0 & 6.82 & 0 & 0 & 0 & 0 \\ 0 & 0 & 6.38 & 0 & 0 & 0 \\ 0 & 0 & 0 & 3.13 & 0 & 0 \\ 0 & 0 & 0 & 0 & 1.8 & 0 \\ 0 & 0 & 0 & 0 & 0 & 1.71 \\ 0 & 0 & 0 & 0 & 0 & 0 \\ 0 & 0 & 0 & 0 & 0 & 0 \end{pmatrix}$$

The first thing to notice is that these are three big-ass matrices! I don't know if it is the coffee I just drank or the size of these bad boys, but this is giving me anxiety! While there is a lot of information to deal with, we'll take it step by step. We can deduce the values for Σ via an eigenvalues problem, meaning that if we multiply U and Σ, we will end up with the projections of all the vectors $\vec{u}$'s into the newly transformed axes; remember that each of the $\vec{u}$ vectors are orthonormal. And, if we constructed Σ in a way such that $\sigma_1 \geq \sigma_2 \geq ... \geq \sigma_n$, then it makes sense to assume that the highest sigma value will be the furthest point projection. This will therefore represent the biggest spread and hence the highest variance. Let's define a measure of energy as:

$$\sum_{i=1}^{n} \sigma_i^2$$

We can then select how much of it to use. Say for example, that we need around 95% of this energy, it then follows that the total energy for our system is equal to:

$$\sum_{i=1}^{6} \sigma_i^2 = 350.30$$

If we wish to use that percentage value, we can remove the values of σ_4, σ_5, and σ_6. Well now, this means two things: we can replace these sigma values with zeroes, which will result in a trio of new U, Σ and V^T matrices, and consequently a new version of A; and if σ_4, σ_5, and σ_6 become zero, then the matrix Σ has the following format:

$$\Sigma = \begin{pmatrix} 15.72 & 0 & 0 & 0 & 0 & 0 \\ 0 & 6.82 & 0 & 0 & 0 & 0 \\ 0 & 0 & 6.38 & 0 & 0 & 0 \\ 0 & 0 & 0 & 0 & 0 & 0 \\ 0 & 0 & 0 & 0 & 0 & 0 \\ 0 & 0 & 0 & 0 & 0 & 0 \\ 0 & 0 & 0 & 0 & 0 & 0 \\ 0 & 0 & 0 & 0 & 0 & 0 \end{pmatrix}$$

Essentially, we are just keeping the information represented by the first three singular values, meaning that we will reduce everything to three dimensions, which in machine learning is called dimensionality reduction. We get rid of some information to make our lives easier. In this example, I wish to represent these relationships visually, as the new contexts are very important. These are the so-called latent variables, which define those cunningly hidden relationships in data. The interpretation of these variables is up to you and it's usually linked to the nature of the problem that you are dealing with. In this case, the context is the music genre. I should point out that there are other applications for dimensionality reduction. For example, clustering algorithms tend to work better in lower dimensionality spaces. Working with a model with all of the available information does not guarantee good results. But, often, if we remove some noise we will improve the capacity of the model to learn.

5.1 Me, Myself and I - The Single Value Decomposition (SVD)

What results from this reduction is a new version of the three amigos U, Σ ,and V^T, which we call the truncated single value decomposition. This version of the decomposition is represented by:

$$\tilde{A}_{m \times n} = U_{m \times t} \cdot \Sigma_{t \times t} \cdot V^T_{t \times n}$$

In this case, t is the number of selected singular values, which for this particular example is:

$$A_{8 \times 6} = U_{8 \times 3} \cdot \Sigma_{3 \times 3} \cdot V^T_{3 \times 6}$$

For the truncated version of U, U_t, we can remove the vectors $\vec{u}_4$ to $\vec{u}_8$ because we replaced σ_4 to σ_6 with zero. The truncated version of U has the following shape and entries:

We can apply the same logic to the matrix Σ because the columns will just be zeros after we replace the single values with zeros. We can therefore remove them:

And V^T will follow:

$$V_t^T = \begin{pmatrix} — & \vec{v_1} & — \\ — & \vec{v_2} & — \\ — & \vec{v_3} & — \\ — & \vec{v_4} & - \\ — & \vec{v_5} & — \\ - & \vec{v_6} & - \end{pmatrix}$$

So, U_t is a matrix of users by context:

$$U_t = \begin{array}{c} \\ user_0 \\ user_1 \\ user_2 \\ user_3 \\ user_4 \\ user_5 \\ user_6 \\ user_7 \end{array} \begin{pmatrix} context_1 & context_2 & context_3 \\ -0.45 & 0.03 & -0.34 \\ -0.35 & 0.17 & 0.70 \\ -0.35 & -0.23 & 0.14 \\ -0.27 & -0.15 & -0.37 \\ -0.28 & 0.51 & 0.13 \\ -0.36 & 0.12 & 0.07 \\ -0.32 & 0.37 & -0.42 \\ -0.38 & -0.68 & 0.09 \end{pmatrix}$$

In this example, the matrix A contains information about user ratings for different songs, so perhaps one interpretation of this context's variables could be the song genre: hip hop, jazz, etc. Now we have another way of relating users, as we go from song ratings to song genre, and we also have a smaller matrix, the matrix U. We can plot this thus:

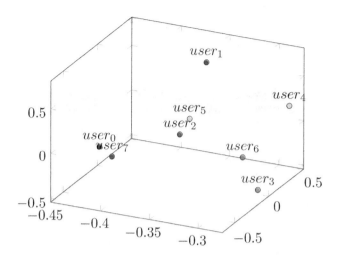

Figure 5.6: The secret space of the *users*.

We can identify some patterns in this plot in terms of similarity of taste based on music genre. Specifically, $user_0$ and $user_7$ seem to have similar tastes and the same seems to be evident with the pairs of $user_2, user_5$ and $user_6, user_3$. It is true that, in this example, you could identify these relationships by simply looking at the matrix A. Still, your matrix will probably have a much higher dimensionality in a real-world situation, and the number of columns will be much larger. With this information, we could segment users based on their music tastes.

An excellent way to do this would be to apply a clustering algorithm to matrix U, as this would return user cohorts who share song preferences. The reason we only use U is that this is a fairly simple example and the singular values are not that distinct.In case of higher dimensionality, with singular values that differ more, it would be better to use $U\Sigma$. The immediate advantages of this include being able to work in a smaller space. Certain algorithms will perform well in these types of space, and it will be computationally cheaper to train them. I won't assume that these concepts are familiar to you, so I will briefly explain. A clustering algorithm has the goal of grouping data that shares identical features, in this case, users that are similar to each other. If you were to run it on matrix U, the outcome would probably be comparable to what I described above. The result would be five groups:

Cluster	Population
1	$user_0$ and $user_7$
2	$user_2$ and $user_5$
3	$user_6$ and $user_3$
4	$user_1$
5	$user_4$

Now, to train a classifier. Hold on one moment; those are some seriously fancy words that circulate in every meeting room and blog on the internet... if you are completely new to machine learning, "training" means to find a configuration for a set number of parameters that characterise a model or an equation.

Okay, back to the point, to train this you need a particular dataset which has labels. These labels are like an outcome that the features will characterise. So, for example, the metrics of the data can be a set of measurements of an engine, and the label can be whether or not the motor was working, a zero for not working and a one otherwise. An algorithm will then accommodate a set of parameters so that those two classes can be separated. For simplicity, say that our model is a line. The equation for this shape is $y = mx + b$. We want to build our model in a way that separates our data into two groups. We aim to have as many zeros below the line as possible, and the maximum number of ones above it. The only thing we can control with that line are the parameters m and b.

So, we will try to find a configuration that can separate our data in the best way possible, and for that, we will be using the provided data set. For example, we can start by randomly drawing a line in space. Our goal is to have all of the one labels above the line and all of the zero labels below the line. As we randomly selected values for m and b, it's likely that these values will not be the best right away but how can we verify this? Well, this is why we need the labels. We can count the number of ones that are above the line and the quantity of zeros that are below the line. We can then tweak the values of m and b to see if this result has improved, meaning that we have more ones above and more zeros below the line respectively:

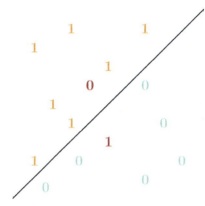

Figure 5.7: An example of separation achieved by a line.

The algorithm I've just described will look randomly for solutions; these types of algorithms are generally called "greedy". It will not be a good heuristic if we don't tell it where to look for the best solutions. But this is just an example; if you are curious, one way of providing an algorithm with information on where to move is via derivatives.

Such concepts will be part of volume two, which will be a book dedicated to calculus. The point here is to understand that, in order to train a model, we need to get a configuration of parameters for an equation (or several) that allows us to achieve our goal as well as we can. So if we have a dataset with labels, there are times when we can better separate data by using latent variables. As we are reducing the number of columns going from A to U, the process is called "dimensionality reduction" in machine learning. This is because we have moved from a higher dimensional space to a lower one, thus reducing the complexity of our problem. Going back to the example of SVD, we haven't looked into the information in matrix V. This particular matrix happens to be an interesting one because it will provide us with the relationship between genre and songs:

$$
V_t = \begin{array}{c} \\ song_0 \\ song_1 \\ song_2 \\ song_3 \\ song_4 \\ song_5 \end{array}
\begin{array}{ccc} context_1 & context_2 & context_3 \\ \left(\begin{array}{ccc} -0.35 & -0.46 & 0.54 \\ -0.27 & 0.45 & -0.30 \\ -0.26 & -0.19 & -0.19 \\ -0.46 & 0.54 & 0.00 \\ -0.35 & 0.27 & 0.60 \\ -0.62 & -0.41 & -0.44 \end{array} \right) \end{array}
$$

Similarly, we can plot this to visually check for closer relationships between songs rather than users:

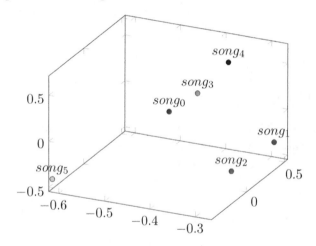

Figure 5.8: Another secret space, the one of the *songs*.

We can see that $song_0$, $song_3$ and $song_4$ seem to be of the same genre, while $song_5$, is probably from a different genre. Finally, $song_2$ and $song_1$ seem to be from the same or a similar genre.

So, we have identified a use case for the single value decomposition. By computing U and V^T, we found new hidden relationships in the data, represented by the latent variables.

We also learned that we could leverage these relationships to train algorithms. Just by itself, this is extremely useful, but we can do more with this decomposition.

If you recall, a matrix is a linear transformation. When you multiply a vector by a matrix, you will be moving these vectors, a movement that can take place within the same space or between different spaces. In this example, we have two matrices that we can use to transform vectors. These are the matrices U and V^T. Let's look into V^T. This matrix has context for columns, which for us is the same as music genre, whereas the rows are songs. So we went from users and song ratings, via matrix A, into a new space, V^T which relates songs by genre. We can create a vector with user ratings and transform it with V^T to understand which genre this particular user is most interested in. Let's check this with an example. Say that we have $user_8$ with the following ratings:

$$user_8 = [5, 0, 0, 0, 0, 0]$$

If we now transform the $user_8$ vector via the matrix V, we will transform $user_8$ into a space of songs by context:

$$[5, 0, 0, 0, 0, 0] \cdot \begin{pmatrix} 0.35 & -0.46 & 0.54 \\ -0.27 & 0.45 & -0.30 \\ -0.26 & -0.19 & -0.19 \\ -0.46 & 0.54 & 0.00 \\ -0.35 & 0.27 & 0.60 \\ -0.62 & -0.41 & -0.44 \end{pmatrix}$$

This results in the following vector:

$$[-1.79, -2.30, 2.74]$$

That new vector shows the new context values or the music genres for $user_8$. With this information, you can, for example, make recommendations. So, you will show this user music from the context or genre 3.

It will also be possible to make a user-to-user recommendation, which recommends music from a user with a preferred
genre that is similar to $user_8$. For this, we need a measure of
similarity. Let's refresh our memory and look at one of the formulas we derived when studying the dot product. If you recall,
there was a way to compute this metric by making use of angles:

$$\vec{a} \cdot \vec{b} = \|a\| \cdot \|b\| \cdot \cos(\theta) \tag{5.14}$$

In this case θ is the angle between the vectors $\vec{a}$ and $\vec{b}$. We
can manipulate equation 5.14 and leave it as:

$$\cos(\theta) = \frac{\vec{a} \cdot \vec{b}}{\|a\| \cdot \|b\|} \tag{5.15}$$

Equation 5.15 is the cosine distance, and the intuition here
is that the smaller the angle between the vectors, the closer they
are to each other. So now you can ask me, why do I need all of
this mathematics if I could simply calculate the cosine distance
between the new $user_8$ and the users in matrix A? Trust me,
I have a severe allergy to work, and if this was not useful, I
wouldn't put it here. Take, for example, users, 3, 4, and 6. If we
calculate the dot product between $user_8$ and these other users
in the space of the matrix A (users by song ratings), we will
find that it is equal to 0. They are therefore not similar at all.
And, users who don't have ratings in common can represent a
problem. To overcome such situations, we can transform $user_8$
into the space of songs by context.

Now, if we calculate the cosine distance between $user_8$ contexts or genre values and the vectors in matrix U, we will get the
user or users that are more similar to $user_8$ in terms of context.
It then follows that:

user	θ
$user_0$	95.80
$user_1$	48.29
$user_2$	29.01
$user_3$	95.10
$user_4$	96.60
$user_5$	67.90
$user_6$	123.50
$user_7$	37.51

We are looking for the smallest angle value, as this dictates the proximity between vectors (and similarity between users). With this in mind, we can see that the users who are more similar to $user_8$ are $user_7$ and $user_2$. We could therefore recommend a highly-rated song from these users to $user_8$ if $user_8$ have not yet rated it. One more useful thing that comes from the single value decomposition equation is that it can represent a matrix as a sum of outer products:

$$A = \sum_{i=1}^{t} \sigma_i \cdot \overrightarrow{u_i} \otimes \overrightarrow{v_i}$$

In this equation t is the number of single values selected for the approximation. Graphically, this equation has the following shape:

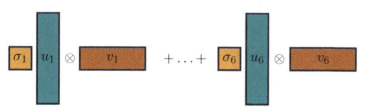

Figure 5.9: A new way to represent a matrix.

This means we can create approximate versions of the matrix A by selecting different numbers of single values, just like the truncated version of the SVDs. Let's see what happens if we choose to approximate A with the first single value. For $\vec{u_1}$, we have the vector:

$$\vec{u_1} = (-0.45, -0.35, -0.35, -0.27, -0.28, -0.36, -0.32, -0.38)^T$$

The vector $\vec{v_1}$ is :

$$\vec{v_1} = (-0.35, -0.27, -0.26, -0.46, -0.35, -0.62)^T$$

Finally, for σ_1, we have the scalar 15.72. An approximation of A with only a single value will be named A_1^*; it can be represented by:

$$A_1^* = \sigma_1 \cdot \vec{u_1} \otimes \vec{v_1}$$

It follows that:

$$A_1^* = \begin{pmatrix} 2.55 & 1.98 & 1.84 & 3.26 & 2.48 & 4.39 \\ 1.98 & 1.54 & 1.43 & 2.54 & 1.93 & 3.42 \\ 1.98 & 1.54 & 1.43 & 2.53 & 1.92 & 3.41 \\ 1.55 & 1.21 & 1.12 & 1.98 & 1.51 & 2.68 \\ 1.64 & 1.27 & 1.18 & 2.09 & 1.59 & 2.82 \\ 2.06 & 1.60 & 1.48 & 2.63 & 2.00 & 3.55 \\ 1.85 & 1.44 & 1.33 & 2.36 & 1.80 & 3.18 \\ 2.19 & 1.70 & 1.58 & 2.80 & 2.13 & 3.78 \end{pmatrix}$$

This is not a very good approximation as A is:

$$A = \begin{pmatrix} 1 & 2 & 3 & 4 & 1 & 5 \\ 4 & 1 & 0 & 3 & 5 & 1 \\ 3 & 0 & 1 & 2 & 2 & 4 \\ 0 & 2 & 1 & 0 & 1 & 5 \\ 0 & 3 & 1 & 3 & 4 & 1 \\ 2 & 1 & 1 & 4 & 2 & 3 \\ 0 & 4 & 1 & 4 & 0 & 3 \\ 5 & 0 & 3 & 0 & 1 & 5 \end{pmatrix}$$

Let's see what happens if we consider two singular values instead of one:

$$A_2^* = \sigma_1 \cdot \overrightarrow{u_1} \otimes \overrightarrow{v_1} + \sigma_2 \cdot \overrightarrow{u_2} \otimes \overrightarrow{v_2}$$

In this particular case, A_2^* seems to be a better approximation for A, which makes sense:

$$A_2^* = \begin{pmatrix} 2.44 & 2.09 & 1.80 & 3.39 & 2.55 & 4.30 \\ 1.43 & 2.10 & 1.19 & 3.20 & 2.28 & 2.92 \\ 2.71 & 0.83 & 1.75 & 1.67 & 1.48 & 4.06 \\ 2.06 & 0.72 & 1.34 & 1.40 & 1.21 & 3.13 \\ 0.04 & 2.85 & 0.49 & 3.98 & 2.57 & 1.39 \\ 1.68 & 1.98 & 1.33 & 3.08 & 2.24 & 3.21 \\ 0.69 & 2.58 & 0.83 & 3.74 & 2.51 & 2.15 \\ 4.35 & -0.40 & 2.52 & 0.28 & 0.82 & 5.71 \end{pmatrix}$$

To verify if this is the case, we will make use of a norm, a measure of distance, called the "Frobenius norm", as in the following formula:

$$\|A\|_F = \sqrt{\sum_{i=1}^{m}\sum_{j=1}^{n} |a_{ij}|^2}$$

151

If we calculate this norm for the differences between the matrix approximations and matrix A, it will provide a good idea of which is closer to A:

$$\|A - A_1^*\|_F = 10.17 \quad \text{and} \quad \|A - A_2^*\|_F = 7.54$$

Why is this useful? Well, one purpose can be to store a vast matrix. For example, suppose you have a matrix of 1 million rows by 1 million columns and you choose to approximate it using the most significant singular value. In that case, it is only necessary to store two vectors of size ten and one scalar (the single value).

Before wrapping up, I would next like to introduce one more technique we mentioned at the start of this book that is widely used in machine learning.

Chapter 6

The Final Stretch - Principal Component Analysis

The analytical technique that will end this journey toward understanding linear algebra is the principal component analysis (PCA). But don't worry, as I bring good news! By now, you know everything that is needed to understand this methodology. In essence, the principal component analysis is a data projection into a new set of axes, or a change of basis that occurs via a linear transformation. Mappings are not a problem for us as we are now experts in manoeuvring vectors through space. So far, we know that when a linear transformation is involved, a matrix is needed, but we still need to understand the goal of this new method to define this linear transformation.

So I will put it out there. The principal component analysis will create a new set of axes called the principal axes, where we will project the data and get these so-called principal components. These are a linear combination of the original features that will be equipped with outstanding characteristics. Characteristics which are not only uncorrelated, but the first components also capture most of the variance in the data, which makes this methodology a good technique for reducing the dimensions of complex data sets. While often seen in a dimensionality reduction context, this technique also has other applications, as these new relationships are latent variables.

Cool, we know what it does and where we can use it, so let's define this linear transformation. The word axes was involved in the explanation, meaning we need orthogonal vectors. By definition if we have a symmetrical positive matrix, the eigenvectors are not only orthogonal but they also have positive and real eigenvalues. Multiplying a matrix by its transpose or the other way around will result in a symmetrical matrix, and with this, we can accommodate a matrix of any size.

Let's pause and check where we are because I am sensing that the desired result is near. We have defined PCA and we've concluded we need to get a new set of axes onto which we can project data. The eigenvectors cover this, but there is one piece missing. These so called components needed to reflect variance. I want to bring to your attention a specific case of matrix multiplication where we multiply A^T by A. Let's think about this operation in terms of dot products. We will end up with a symmetrical matrix where the non-diagonal entries represent how much the rows relate to the columns.

With this, we are one step away from a covariance matrix. The numbers in this particular matrix reflect how much variables vary with each other. The only thing we're missing is a way to centre the data around the mean. Given that the covariance matrix is symmetric positive, its eigenvalues are positive and the eigenvectors are perpendicular. Considering that the eigenvalues represent a scaling factor that comes from a covariance matrix, the largest value will correspond to the highest direction of variance; therefore, the correspondent eigenvector will be the first principal component.

We can apply this same logic to each of the other vectors and then order them. The result will be the principal axes on which we can project the data to obtain the principal components. I know that's a lot of words, but no numbers, and no equations! That's not our style! So let's consider an example, a fabricated one, but still, it will help us comprehend this technique. Say that we have the following set of data:

user	totalBetsValue	totalWon	totalDaysPlayed	averageBetSize	totalSessions
1	3000	0	4	30	4
2	10453	0	1	100	1
3	21500	4230	6	50	7
4	10000	2000	12	10	14
5	340	10	10	1	10
6	5430	2000	4	5	70
7	43200	4320	10	4	32
8	2450	100	8	5	12

The data can represent, for example, the features of users for some online casino games. Our goal is to transform this data with principal component analysis, so the first step is to calculate a covariance matrix. For that, we need to centre the data. To do this, we can use the following equation:

$$x^* = \frac{x - \bar{x_j}}{\sigma_j} \qquad (6.1)$$

This means that, for each element in column j, we will subtract the mean of the same column and then divide the results by the standard deviation of this same column. If we do this to each column, we have standardised the data. So let's start by computing each column's mean and standard deviation:

Metric	totalBetsValue	totalWon	totalDaysPlayed	averageBetSize	totalSessions
$\bar{x_j}$	12046.62	1582.50	6.87	25.62	18.75
σ_j	13329.42	1751.10	3.51	32.23	21.264

Applying equation 6.1, the table with standardised data is as follows:

The Final Stretch - Principal Component Analysis

user	totalBetsValue,	totalWon ,	totalDaysPlayed,	averageBetSize ,	totalSessions ,
1	-0.68	-0.90	-0.82	0.14	-0.69
2	-0.12	-0.90	-1.67	2.31	-0.83
3	0.71	1.51	-0.25	0.76	-0.55
4	-0.15	0.24	1.46	-0.48	-0.22
5	-0.88	-0.90	0.89	-0.76	-0.41
6	-0.50	0.24	-0.82	-0.64	2.41
7	2.34	1.56	0.89	-0.67	0.62
8	-0.72	-0.85	0.32	-0.64	-0.32

We are now ready for linear algebra, so let's go from that ugly table into a more familiar format in which to display data, a matrix. And that is right, it will be good old matrix A! So let A be such that:

$$A = \begin{pmatrix} -0.68 & -0.90 & -0.82 & 0.14 & -0.69 \\ -0.12 & -0.90 & -1.67 & 2.31 & -0.83 \\ 0.71 & 1.51 & -0.25 & 0.76 & -0.55 \\ -0.15 & 0.24 & 1.46 & -0.48 & -0.22 \\ -0.88 & -0.90 & 0.89 & -0.76 & -0.41 \\ -0.50 & 0.24 & -0.82 & -0.64 & 2.41 \\ 2.34 & 1.56 & 0.89 & -0.67 & 0.62 \\ -0.72 & -0.85 & 0.32 & -0.64 & -0.32 \end{pmatrix}$$

So A^T is:

$$A^T = \begin{pmatrix} -0.68 & -0.12 & 0.71 & -0.15 & -0.88 & -0.50 & 2.34 & -0.72 \\ -0.90 & -0.90 & 1.51 & 0.24 & -0.90 & 0.24 & 1.56 & -0.85 \\ -0.82 & -1.67 & -0.25 & 1.46 & 0.89 & -0.82 & 0.89 & 0.32 \\ 0.14 & 2.31 & 0.76 & -0.48 & -0.76 & -0.64 & -0.67 & -0.64 \\ -0.69 & -0.83 & -0.55 & -0.22 & -0.41 & 2.41 & 0.62 & -0.32 \end{pmatrix}$$

One more step, and we will have the desired covariance matrix. The only thing we're missing is financial freedom, oh, oops... Sorry, my mind drifts away sometimes. I meant to say, we need to multiply A^T by A and then divide by the number of entries. Shall we call the resultant of this operation matrix M?

$$\frac{A^T A}{8} = M = \begin{pmatrix} 1.00 & 0.84 & 0.23 & 0.02 & 0.13 \\ 0.84 & 1.00 & 0.29 & -0.14 & 0.33 \\ 0.23 & 0.29 & 1.00 & -0.73 & -0.01 \\ 0.02 & -0.14 & -0.73 & 1.00 & -0.47 \\ 0.13 & 0.33 & -0.01 & -0.47 & 1.00 \end{pmatrix}$$

So as expected, M is a symmetrical matrix. Where are the eigenvectors? There are several ways to get them. One is to use the eigen decomposition, so let's start by exploring that. This numerical technique will return three matrices, two of which will have what we are after, the eigenvalues and the eigenvectors. This time I will use a computer to calculate the eigenvectors and eigenvalues. We are looking for a representation of M like:

$$M = P\Sigma P^{-1}$$

Where P is:

$$P = \begin{pmatrix} 0.45 & 0.53 & 0.44 & -0.43 & 0.34 \\ 0.53 & 0.42 & -0.38 & 0.60 & -0.13 \\ -0.14 & 0.01 & -0.54 & -0.04 & 0.82 \\ 0.69 & -0.68 & -0.11 & -0.19 & 0.05 \\ 0.00 & -0.24 & 0.58 & 0.64 & 0.42 \end{pmatrix}$$

And:

$$\Sigma = \begin{pmatrix} 2.29 & 0 & 0 & 0 & 0 \\ 0 & 1.48 & 0 & 0 & 0 \\ 0 & 0 & 1.02 & 0 & 0 \\ 0 & 0 & 0 & 0.14 & 0 \\ 0 & 0 & 0 & 0 & 0.08 \end{pmatrix}$$

Right, so P is a matrix with the eigenvectors, and this will be where we find our principal axes. It happens to be already sorted by eigenvalue magnitude. On the other hand, in the matrix Σ we have all the eigenvalues. These represent what can be called the explainability of the variance, how much of the variance present in the data is "captured" by each component. This means we can derive excellent insights from these magnitudes when we want to apply principal component analysis as a dimensionality reduction technique. For example, we could percentage the eigenvalues and then understand how much variance they each explain:

$$\sum_{i=1}^{5} \lambda_i = 5 \tag{6.2}$$

We have five eigenvalues, and we can sum them using 6.2, where λ_i represents the eigenvalues. Now, by dividing each of them by five, we find the individual percentage of variance that each distinct eigenvalue explains.

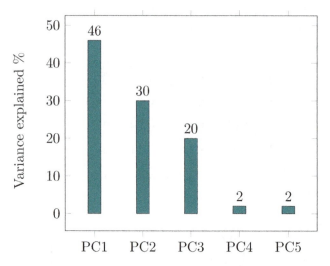

Figure 6.1: The amount of information captured by each component.

We can see that two components, namely PC1 and PC2 explain 76% of the variability. Considering the case of dimensionality reduction, let's pick the two largest eigenvalues and transform our original data set to create the first two components for each user in the data set. Choosing the first two components will reduce the columns of P. We will call this new version P_t, and it has the following format:

$$P_t = \begin{pmatrix} 0.45 & 0.53 \\ 0.53 & 0.42 \\ -0.14 & 0.01 \\ 0.69 & -0.68 \\ 0.00 & -0.24 \end{pmatrix}$$

The linear combinations that represent the first two components are described by the following two equations:

$$\begin{aligned} PC1 = 0.45 \cdot \text{totalBetsValue} + 0.53 \cdot \text{totalWon} \\ -0.14 \cdot \text{totalDaysPlayed} + 0.69 \cdot \text{averageBetSize} \\ +0.00 \cdot \text{totalSessions} \end{aligned} \quad (6.3)$$

$$\begin{aligned} PC2 = 0.53 \cdot \text{totalBetsValue} + 0.42 \cdot \text{totalWon} \\ +0.01 \cdot \text{totalDaysPlayed} - 0.68 \cdot \text{averageBetSize} \\ -0.24 \cdot \text{totalSessions} \end{aligned} \quad (6.4)$$

The only thing missing is to project the data into this new space. We need to transform our versions of the scaled data with the matrix P_t. For that, we can use the following equation:

$$M_{reduced} = M_{8\times5} \cdot P_{5\times2}^T \quad (6.5)$$

By performing the matrix multiplication in 6.5, we will create the first and second components for all users in the features set:

user	PC1	PC2
1	-1.46	-0.26
2	-2.56	1.69
3	0.51	1.65
4	0.84	-0.81
5	-0.30	-1.60
6	0.63	-0.55
7	2.81	1.10
8	-0.48	-1.22

These components are also called latent or hidden variables; relationships that are hidden in the data and are the result of linear combinations. It is possible to give these some meaning (a way of interpretation), and for that, we can use the expressions that define them, as shown in 6.3 and 6.4 that define them.

I want to point out that this step is dependent on the business problem and what you think makes most sense. This is probably one of the only times when mathematics is similar to poetry. Some will disagree with me and say that equations are pure poetry, and whilst I may not have reached that level of sensitivity, I do wonder what it's like to live with that state of mind. Maybe it is excellent.

Anyway, back to our example. So, let's say that principal component number one could be intensity. The higher positive coefficients in equation 6.3 are those that correspond to totalBets, totalWon, and averageBetSize. The second principal component is harder to attribute a meaning to, but risk exposure could make sense as these players have significant losses:

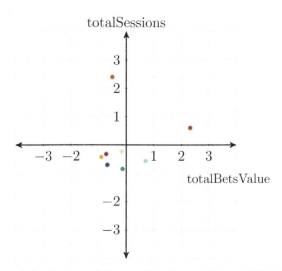

Figure 6.2: A view of the totalSession and the totalBetsValue.

Seeing as we reduced the data from five dimensions to two and we attributed some meaning to the new components, we can plot these new coordinates and see if anything interesting comes from it:

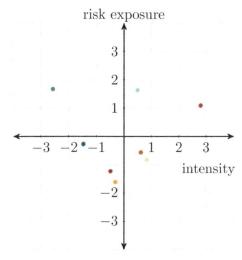

Figure 6.3: A peak into the secret space! First two components.

This is somehow an interesting perspective in the context of gambling. Some players like to risk their balance but play in a less intense session. We can also observe a case of a player that enjoys risk and intensity. These hidden variables are excellent indicators for situations where human emotion is involved, as they can capture some of that temperament.

There are other methods that can be leveraged to get to the principal components. While we have so far explored the eigen decomposition, it is also possible to calculate these metrics via the single value decomposition. We defined M as a covariance matrix while A was the matrix with entries that are a standardised version of the data in the original features table. So, M can be defined by the following equation:

$$M = \frac{A^T \cdot A}{N} \tag{6.6}$$

In equation 6.6, N is the number of rows in matrix A. Now, say that, instead of performing an eigen decomposition on M, we choose to use the single value decomposition and we do it on matrix A. It follows that:

$$A = U \cdot \Sigma \cdot V^T$$

Substituting this information into equation 6.6 it follows that:

$$M = \frac{(U \cdot \Sigma \cdot V^T)^T (U \cdot \Sigma \cdot V^T)}{N}$$

We can manipulate that equation a little bit:

$$M = \frac{V \cdot \Sigma^T \cdot U^T \cdot U \cdot \Sigma \cdot V^T}{N}$$

Finally:

$$M = \frac{V \cdot \Sigma^2 \cdot V^T}{N} \tag{6.7}$$

From here, we can conclude that the right singular vectors are the principal directions. Following the same logic, just as we did with the eigen decomposition, we will obtain the principal components by:

$$A.V$$

And A is equal to $U \cdot \Sigma \cdot V^T$, so:

$$A \cdot V = U \cdot \Sigma \cdot V^T \cdot V$$

Which means that:

$$A \cdot V = U \cdot \Sigma$$

So the principal components are given by $U \cdot \Sigma$ where the entries of Σ are calculated by:

$$\lambda_i = \frac{\sigma_i^2}{N}$$

6.1 We Did It!

And for linear algebra, that is it! It was a bit of a journey for me and the vectors, and hopefully a pleasant one for you. The start was all about a simple oriented arrow. From there, we deduced a lot of mathematics to finally arrive at two techniques that have abundant applications in machine learning: the principal component analysis and the single value decomposition. To get here, we went over plenty of notation and equations that equipped us with what we needed to learn or refine new concepts: these included dot products, linear transformations, the determinant, and the eigenvectors, to name a few.

As this book is about linear algebra, it would be gratifying if, by now, you felt comfortable with this field of mathematics, but more importantly, my hope is that this book may have made you think, even if just for a little bit. Thinking can be the stairway to heaven. Well, perhaps it would be if you added psychedelics to the experience. But I will leave this to your discretion, stating that I only encourage the reading of mathematics and that this will be the only unproven concept in the book, so don't lawyer up!

What I do know is that thinking is a gate; once you got through this gate, it will make you join the path of knowledge, and with this comes freedom. In the past, it was common to state that the truth would set you free. But, like many things, this has changed. It seems now that freedom is unfortunately linked to capital, and the sentence "I am not money-driven", that for so long was almost the standard line you had to use if you wished to be hired by a corporation, has probably lost its meaning.

These days, the idea of spending money that we don't have has been normalised. With this comes a dependency on a system that is controlled by people who, on many occasions, have made it clear that they are only concerned with buying themselves a bigger boat. Think about it, three presses on a button on some computer that belongs to an institution that likely has central or federal in its name, and overnight, you just got fucked. It takes the same amount of time for you to order a juicy burger on your phone.

The edge is knowledge, and mathematics will help you get there.

Peace

Jorge

Please subscribe to the mailing list if you want to receive an email when the subsequent volumes are released.

Subscribe here

Made in the USA
Las Vegas, NV
09 May 2024

89733376R00095